DEMOLITION

DEMOLITION

THE ART OF DEMOLISHING, DISMANTLING, IMPLODING, TOPPLING & RAZING

BY HELENE LISS
WITH THE LOIZEAUX FAMILY OF CONTROLLED DEMOLITION, INC.

BLACK DOG
& LEVENTHAL
PUBLISHERS
NEW YORK

PUBLISHED BY BLACK DOG & LEVENTHAL PUBLISHERS, INC.
151 WEST 19TH STREET, NEW YORK, NY 10011

DISTRIBUTED BY WORKMAN PUBLISHING COMPANY
708 BROADWAY, NEW YORK, NY 10003

PRINTED IN SPAIN BY ARTES GRÁFICAS TOLEDO S.A.U.
D.L. TO: 1377-2000

DESIGNED BY 27.12 DESIGN, LTD.

ISBN: 1-57912-149-7

H G F E D C B A

LIBRARY OF CONGRESS CATALOGING-IN-PUBLICATION DATA

LISS, HELENE
DEMOLITION : THE ART OF DEMOLISHING, DISMANTLING, IMPLODING, TOPPLING & RAZING / BY HELENE LIS
WITH THE LOIZEAUX FAMILY OF CONTROLLED DEMOLITION, INC.
P. CM.
ISBN 1-57912-149-7
1. WRECKING I. CONTROLLED DEMOLITION. II. TITLE.

TH447 .L57 2000
690'.26--DC21

00-044482

ACKNOWLEDGMENTS

Writing this book has been a fabulous adventure, made possible only becauseof the help, support, and generosity of many individuals to whom I am indebted:

...Mark, Doug and Stacey Loizeaux, as well as the entire extended CDI family, who welcomed me so warmly into their world. It has truly been an honor to get to know and work with such an incredible family.

...Mike Taylor, Executive Director of the National Association of Demolition Contractors; Uwe Kausch and Parnell Thill of LaBounty Manufacturing; Bill Moore of Brandenburg Industrial Service Company; Sharon Holling and Jan-Willem van den Brand of Caterpillar; Ron Dokell of Demolition Management Consultants; Doug Cool and Steve Aman of Aman Environmental Construction, Inc.; Dan Hoffman of Asset Recovery Contracting; and Art Dore of Dore & Associates Contracting, Inc.; all of whom shared generously of their knowledge and expertise about this fascinating industry.

...Pete Pedersen and Amanda Arto of Edelman Worldwide, who took me under their wing in Seattle.

... All of my friends, family and colleagues who indulged me whenever I wanted to "talk demolition."

... Michael Driscoll, Pamela Horn and the staff of Black Dog & Leventhal, who helped turn the vision of this book into reality.

... JP Leventhal, who presented me with this opportunity, as well as a shot at my first fifteen minutes of fame.

... And, finally, my parents, who instilled in me a love of challenge and the drive to succeed.

TABLE OF CONTENTS

INTRODUCTION 9

    DEMOLITION AND POLITICAL CHANGE 10

    DEMOLITION AS OFFENSE 12

THE PROCESS OF DEMOLITION 14

DEMOLITION EQUIPMENT 18

    WRECKING BALL 19

    EXCAVATORS 20

    FRONT-END LOADERS 22

    ATTACHMENTS 24

        SHEARS 26

        ON GRADE PROCESSORS 27

        GRAPPLES 27

        UNIVERSAL PROCESSORS 27

        CONCRETE PULVERIZERS 27

CONVENTIONAL DEMOLITION 28

    EBBETTS FIELD 30

    POLO GROUNDS 31

    PENNSYLVANIA STATION 32

    SEARS CATALOGUE WAREHOUSE CENTER 34

    A8 HIGHWAY BRIDGE 35

    MILLER PARK STADIUM 36

THE EVOLUTION OF IMPLOSION 38

    THE LOIZEAUX FAMILY
    OF CONTROLLED DEMOLITION, INC. 39

HISTORY OF EXPLOSIVES 42

    BLACK POWDER 42

    NITROGLYCERIN 42

    DETONATING CORD 43

    DELAY SYSTEMS 43

    SHAPED CHARGES 43

    BLASTING MACHINES 43

THE PROCESS OF IMPLOSION 44

    LOADING THE BUILDING 44

    HOW MUCH EXPLOSIVES TO USE? 44

    WHERE DOES IT ALL GO? 45

    WHAT ABOUT THE AIR? 45

    CAN ANYTHING BE IMPLODED? 45

    HOW IT ALL FALLS TOGETHER 45

"THAT WHICH GOES UP, SHALL COME DOWN" 46

    MENDES CALDIERA BUILDING 46

    TRAVELER'S BUILDING 48

    ARDLER MSD FLATS 50

    J.L HUDSON DEPARTMENT STORE 52

    LARGE PHASED ARRAY (LPAR) FACILITY 54

    ANZ BANK 56

    BARBADOS HILTON HOTEL 56

    PARKVIEW HILTON HOTEL 58

    MONTLUCON COMPLEX 59

TABLE OF CONTENTS

BILTMORE HOTEL — 60

OMEGA TOWER — 62

NAM SAN FOREIGNERS' APARTMENT COMPLEX — 64

UN PAVILION — 65

HOLIDAY INN — 66

SOUTHWARK TOWERS — 68

LAS VEGAS — 70

SANDS — 71

HACIENDA HOTEL — 72

ALADDIN HOTEL — 74

DUNES HOTEL NORTH TOWER — 76

HOLLYWOOD — 78

DR PEPPER BUILDING — 78

ORLANDO CITY HALL — 80

TRAYMORE HOTEL — 83

LANDMARK HOTEL — 84

BRIDGES — 86

GEORGE P. COLEMAN BRIDGE — 86

SUNSHINE SKYWAY BRIDGE — 88

MUSCATINE BRIDGE — 91

MANCHESTER BRIDGE — 91

COURT STREET BRIDGE — 93

TALMADGE MEMORIAL BRIDGE — 94

PARKERSBURG-BELPRE SUSPENSION BRIDGE — 95

CHIMNEYS — 96

KENNECOTT COPPER SMELTER STACK — 96

HARVARD MEDICAL CHIMNEYS — 97

INDUSTRIAL STRUCTURES — 98

U.S. STEEL FURNACE — 98

SHARON STEEL PLANT STRUCTURES — 100

PSE&G GAS HOLDER — 101

MADRAS COOLING TOWERS — 102

SEQUEDIN COOLING TOWER — 103

SPORTS ARENAS — 104

OMNI ARENA — 104

ST. LOUIS ARENA — 106

KINGDOME — 108

MILITARY EQUIPMENT DEMOLITION — 112

SCUD MISSILE AND RELATED EQUIPMENT DISPOSAL — 112

FINISHING NATURE'S WORK — 114

COLONIA ROMA — 114

MARINA BUILDING — 116

EMERGENCY RESPONSE — 118

SHEIK ABDULLAH ALAKL RESIDENTIAL & COMMERCIAL CENTER — 118

MURRAH FEDERAL BUILDING — 120

BIBLIOGRAPHY — 122

PHOTOGRAPHY CREDITS — 123

INDEX — 124

INTRODUCTION

Observe any child with a set of blocks and it becomes obvious that the desire to build is innate. But another thing also becomes obvious—that the desire to knock down is innate as well. In fact, it is often difficult to tell which act the child enjoys more.

Since the dawn of time, man has striven to build—to leave his mark on the landscape, to improve his standard of living, to provide shelter for his family. And since not long after this dawn, man has sought also to demolish structures that for one reason or another no longer suited his needs. Demolition of this type is often viewed as an end, but more accurately it should be viewed as a beginning: Demolition of the old or outdated allows the physical constructs of a developing society to evolve.

DEMOLITION AND POLITICAL CHANGE

Often, demolition is the act that follows a significant change in politics or thought. In many ancient civilizations, the ascension of a new ruler or regime was followed by the destruction of structures associated with the predecessor. Oftentimes these materials were recycled for use in structures constructed by the new government.

Goddesses of demolition, of destruction and renewal, existed in the pantheons of many ancient cultures—the Egyptian Sekhmet, Indian Kali, Irish Morrigan, and Scottish Cailleach. The coupling of Egypt's lion-headed goddess Sekhmet, with Ptah, the Master Mason, is not viewed as coincidental—that which is destroyed by Sekhmet is rebuilt by Ptah. The union of these two deities produced Imhotep, the god of healing and medicine. Imhotep was the architect of the step pyramid complex of King Djoser (2630-2611 B.C.) at Sakkara.

We can look to more recent history for evidence of demolition as indicative of change: The storming, and demolition, of the Bastille in Paris was the symbolic start of the French Revolution. The Bastille, considered a symbol of the despotism of the reigning Bourbons, was an imposing structure—eight towers 100 feet high, linked by walls of equal height and surrounded by a moat more than 80 feet wide. Though originally built as a fortress in the Middle Ages, the Bastille came to be used as a state prison in the 17th century. On average, the Bastille housed 40 prisoners interned by lettre de cachet or direct order of the king, from which there was neither appeal nor recourse. Many of these individuals were political troublemakers and individuals held at the request of their families, often in an effort to coerce obedience or prevent marring of the family's name. (Along with prisoners, the Bastille also housed banned books.)

On the morning of July 14, 1789, an angry mob of Parisians stormed and captured the Bastille, in a dramatic action that triggered the end of the Bourbon regime. The Bastille was subsequently demolished by order of the Revolutionary government. In 1880 the anniversary of the storming was made into a French national holiday; Bastille Day is celebrated with parades, speeches, fireworks, and slogans such as A bas la Bastille! ("Down with the Bastille!")

For an even more recent example of demolition as a step toward renewal, we need only look to the destruction of the Berlin Wall in November 1989. After the end of World War II, Berlin was completely surrounded by Soviet-occupied territory. This territory officially became the country of East Germany in 1949. Berlin itself was divided—West Berlin, occupied by the American, British and French, was supported by the Federal Republic of Germany (commonly known as West Germany). East Berlin was occupied by the Soviets.

East Germany tried to slow the exodus of its citizens to the west, by partitioning the city; this political division became physical on August 13, 1961. Residents of Berlin awoke to a barbed-wire barrier built by East German soldiers and militia, which was soon replaced by a concrete wall 12 feet high and 103 miles long. Tank traps and ditches were

Left: Sekhmet, Egyptian goddess of destruction.

Above: The Storming of the Bastille.

y citizens (above) helped demolition experts (right) bring down the Berlin Wall in 1989.

built along the east side of the Wall, which had only two openings, both closely guarded.

Though the Wall was protected by mines, attack dogs and guards with shoot-to-kill orders, at least 2.7 million people tried to escape—by climbing, vaulting, tunneling, crashing through checkpoints, swimming through canals or stowing away in ships transporting cargo across the border. Between 400 and 800 individuals lost their lives in attempts to cross the border.

The Wall lost its power in the summer of 1989 when Hungary announced that it would allow East Germans to pass through on their way to Austria and West Germany. When the news spread that the restrictions on travel in either direction would be lifted, private citizens took to the Wall and began demolishing entire sections without interference from government officials.

East Germany eventually participated in the destruction of the Wall and reunited with West Germany in 1990 as one nation—The Federal Republic of Germany.

"It was still very cold and dark at 5 AM. Perhaps 7,000 people were pressed together, shouting, cheering, clapping. We pushed through the crowd. From the East German side we could hear the sound of heavy machines. With a giant drill, they were punching holes in the wall. Every time a drill poked through, everyone cheered. The banks of klieg lights would come on. People shot off fireworks and emergency flares and rescue rockets. Many were using hammers to chip away at the wall. There were countless holes. At one place, a crowd of East German soldiers looked through a narrow hole. We reached through and shook hands. They couldn't see the crowd, so they asked us what was going on and we described the scene for them. Someone lent me a hammer and I knocked chunks of rubble from the wall, dropping several handfuls into my pocket. The wall was made of cheap, brittle concrete: the Russians had used too much sand and water."

— Adreas Ramos, witness to the wall's demolition.

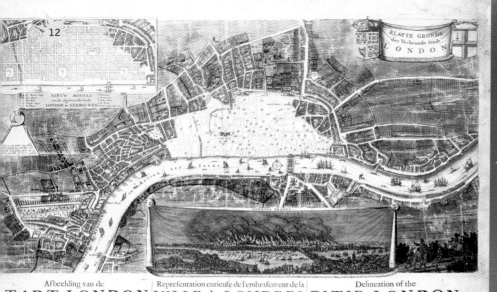

Afbeelding van de
TADT LONDON.
Aeuwijzende hoe verre de zélve verbrandt is, en wat
plaetfen noch overgebleven zijn.

Reprefentation curieufe de l'embrafement de la
VILLE de LONDRES.
Avec une Demonftration exacte de ce qui en eft
demoré de refte.

Delineation of the
CITIE LONDON,
Shewing how far the faid citie is burnt down, and what
places doe yet remain ftanding.

DEMOLITION AS OFFENSE

The Great Fire of London, which raged from September 2-5, 1666, was the worst fire in London's history. It destroyed a large part of the City of London, including most of the civic buildings, the old St. Paul's Cathedral, 87 parish churches and about 13,000 houses. Further devastation was avoided by an early example of controlled destruction.

The fire began accidentally in the house of the king's baker, near London Bridge. Legend has it that the baker left an ember burning in the kitchen—essentially, he left the oven on. A violent east wind encouraged the fire, which was extinguished on Thursday, only to burst forth again that evening. In an act of offense as defense, King Charles II ordered a slew of houses demolished through the use of gunpowder in order to starve the fire. His plan worked, and the fire was finally mastered.

The Great Fire is commemorated with a column, "The Monument," erected in the 1670s near the source of the fire.

Demolition was also used to curb the path of a fire in San Francisco. A great earthquake struck San Francisco on April 18, 1906, at 5:13 AM, and it was followed by a tremendous fire that burned for four days. In an attempt to build a firebreak, Colonel Charles Morris of the Artillery Corps ordered his army to demolish mansions along Van Ness Avenue. It is believed that more than 3,000 individuals died as a result of the earthquake and subsequent fire. Damage was estimated at $500,000,000 in 1906 dollars.

Top: In this 17th century map of London, the white area represents the portion of the city wiped-out by the Great Fire. Preventative demolition spared the rest of the city.

Right: The Monument to the Great Fire of London

Opposite page: San Francisco was devastated by fire in 1906 (left), as spectators looked on from a nearby hill (right). Many homes were intentionally demolished to keep the fire from spreading further.

THE PROCESS OF DEMOLITION

Once the decision is made to demolish a structure, demolition contractors are invited to bid on the job. Many factors play a role in the determination of the bid: What's the nature of the structure? How was it constructed, and what condition is it in? How hazardous is it, and what types of insurance will be needed? What's the best way to bring it down? What types and how many pieces of machinery will be needed? How many workers, and for how long? How much material can be salvaged for scrap? Often a prime contractor will be selected, and that contractor will then subcontract aspects of the job (asbestos removal, implosion) to other contractors.

Though not their most profitable projects, the demolition of single-family residences is a fairly common practice for most demolition contractors. Sometimes homes are brought down because they've been damaged in fires or earthquakes; other times they are brought down to clear land for other kinds of construction projects. Using a tracked front-end loader, most homes can be demolished in just one hour. The next step involves sorting the rubble, which is then hauled on debris trucks for recycling or disposal.

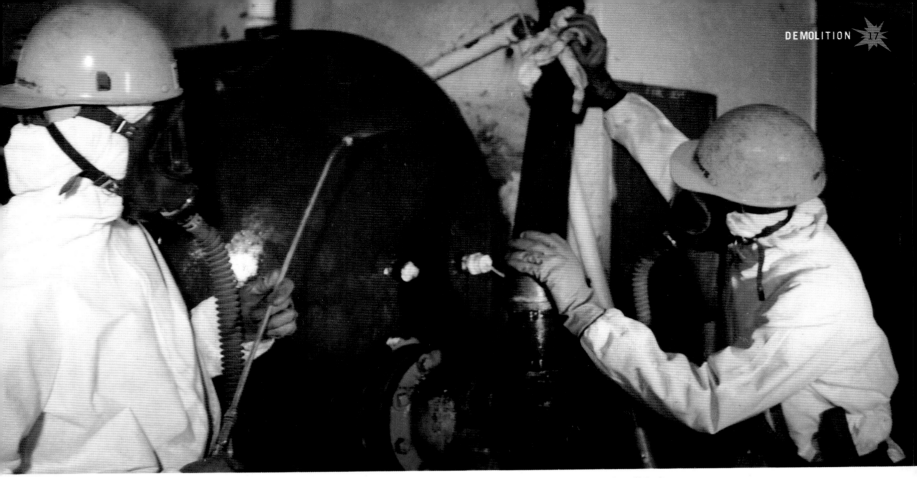

Per regulations, workers must remove toxic substances (like asbestos, which is being removed above) before a building can be demolished.

Following are some common misconceptions, taken from "10 Common Misconceptions about the Demolition Industry" produced by the National Association of Demolition Contractors. Structure, amount of salvage it will yield, and manner of demolition employed are all factors that influence the cost of a particular job.

DEMOLITION CONTRACTORS PRIMARILY IMPLODE BUILDINGS.
The reality is that implosion accounts for less than 1% of all demolition work—the other 99% is handled with specialized heavy demolition equipment or skilled manual techniques. Hand labor and tools, such as sledgehammers, picks, wrecking bars, shovels and steel cutting torches are frequently used in locations that are inaccessible by mobile equipment or in the process of renovation.

DEMOLITION CONTRACTORS DESTROY MANY STRUCTURES THAT SHOULD BE SAVED.
In some cases demolition contractors, through the process of selective demolition, are able to help preserve historic structures. In other cases, building interiors can be gutted and renovated without disturbing the façade or other important architectural details. Demolition contractors contribute to the overall aesthetic quality of communities by removing deteriorated roads, bridges, and unstable structures damaged by fire, earthquake, or weather.

DEMOLITION CONTRACTORS DON'T PARTICIPATE IN THE NATION'S RECYCLING EFFORT.
The demolition industry's tradition of salvaging building elements and materials for reuse predates the national recycling movement. Recycling is a major part of the demolition contractor's business and represents 20 to 50% of some companies' revenues.

- Scrap iron, "rebar" (reinforced rods in concrete), aluminum, stainless steel and copper are typically recycled
- Wood is often reused as building lumber, landscape mulch, pulp chips and fuel
- Bricks are cleaned and reused
- Concrete debris is pulverized and used as road sub-base and fill material
- Doors, sinks, bathtubs and used building materials are resold

DEMOLITION IS DANGEROUS.
While demolition has the potential to create unsafe work conditions, contractors have designed programs to prevent both on- and off-site accidents. Many demolition contractors employ in-house safety directors to prepare written standards, conduct safety meetings and provide employees with the most advanced safety equipment. Government agencies, including the Department of Transportation, Environmental Protection Agency and Occupational of Safety and Health Administration, have further helped shine the spotlight on safety.

DEMOLITION METHODS NEVER CHANGE.
Demolition practices are constantly evolving, becoming faster, safer and more cost-effective. In days past the basic tools of demolition were crane, wrecking ball and front-end loader; today, standard equipment includes the hydraulic excavator with its many attachments—grapples, shears, hammers and concrete crushers.

DEMOLITION IS AN UNSOPHISTICATED BUSINESS.
Because they must often make decisions involving mechanical and electrical systems, engineering, and environmental regulations, demolition professionals must have a working knowledge of both construction and the law. Additionally, trained professionals are required to identify, remove and dispose of any and all hazardous, toxic or regulated materials.

ALL DEMOLITION CONTRACTORS ARE BASICALLY THE SAME.
Demolition contractors are like doctors—expertise varies with experience and investment in training and equipment. There are residential, commercial and industrial demolition contractors whose specialties include highway demolition, excavating and foundation removal, selective structural demolition, interior strip-outs, blasting, environmental remediation and equipment removal.

DEMOLITION IS EXPENSIVE.
Commercial demolition work generally costs less than 2% of the replacement cost of the building. The age of the structure, amount of salvage it will yield, and manner of demolition employed are all factors that influence the cost of a particular job.

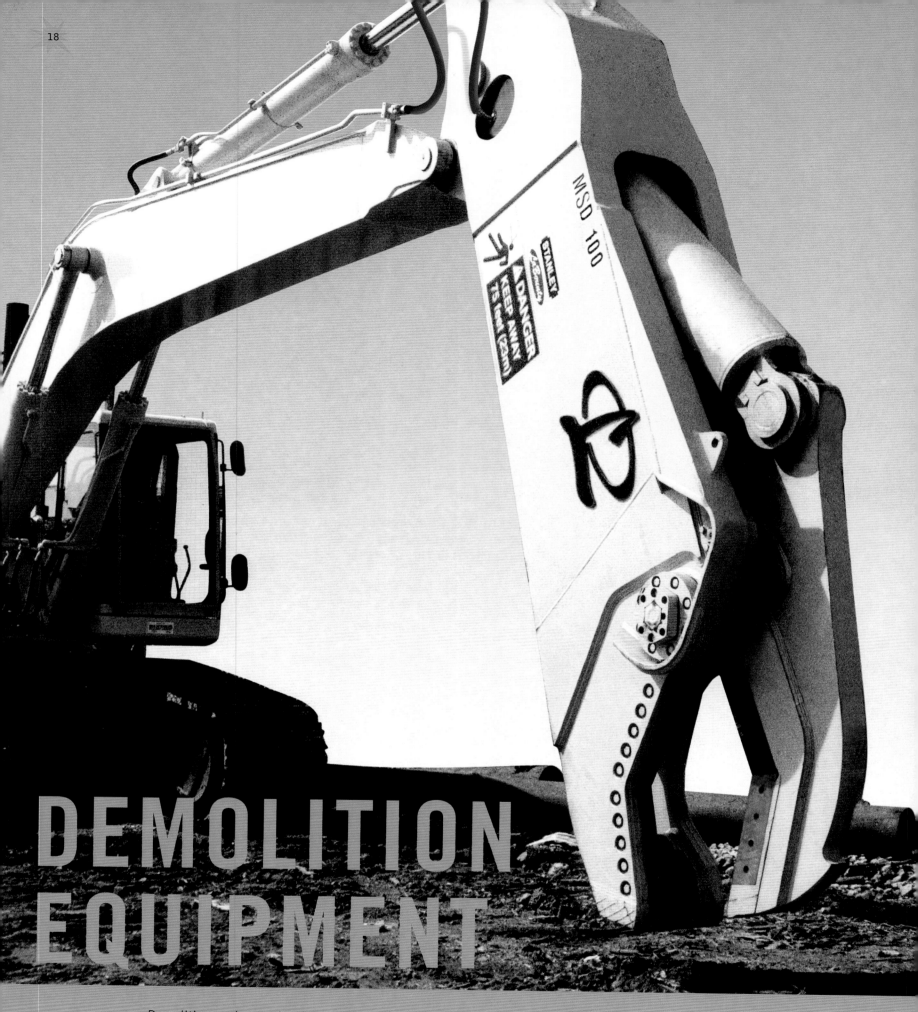

DEMOLITION EQUIPMENT

Demolition equipment has been revolutionized through three technological innovations: the development of hydraulics, the addition of computer controls and the development of stick-end attachments for excavators. Together, these innovations have introduced an element of speed, safety and economy previously unavailable.

Above: A wrecking ball at work in Kansas City, Missouri.

Wrecking Ball

Upon hearing the word "demolition," most people immediately conjure an image of a wrecking ball. This is not surprising, when one considers that the wrecking ball is the most popular technique for demolishing buildings of four or more stories. Though fairly quick and dramatic to watch, wrecking balls may not be the easiest way to bring down all structures of this size; in some instances, they are being replaced with backhoes armed with triple booms, nibblers and an 80-foot reach.

It is believed that the wrecking ball evolved from the battering ram, a popular medieval weapon consisting of a heavy log with a metal knob or point at the front. Typically, these tools were used to batter down the gates or walls of a besieged city or castle. The ram was suspended by ropes from the roof of a movable shed and swung back and forth against the structure by the ram's operators. The roof of the shed, from which the operators worked, was usually covered with animal skins in an attempt to protect the operators from stones or fiery materials lobbed by the enemy.

Left: Though wrecking balls are no more than large iron weights suspended from a crane, they are far from simple. Operating a "crane-and-ball" isn't as easy as it appears—the operator must be vigilant to make sure the ball doesn't bounce back and hit the crane's boom.

Opposite page: A massive shear attachment can take apart a building piece by piece.

Above and right: Excavators at work

Excavators

Excavators evolved from steam shovels, first using internal combustion engines, then employing hydraulics in the 1960s. The power of the excavator lies in its humanlike arm that can reach over 40 feet and crush concrete. The excavator is incredibly versatile because of the multitude of attachments that can be placed on the end of the arm like shears that cut, grapples that grab, and breakers that shatter. Skilled operators, using the excavator's sophisticated, computer-controlled hydraulics, can easily manipulate these attachments while applying 30,000 pounds of force. Excavators are also extremely proficient in removing rebar from rubble, which can be a tedious and difficult task. (Rebar is short for "reinforcing bar," and refers to a steel rod with ridges used to strengthening concrete.)

Excavators, when compared to other pieces of heavy equipment, have relatively small engines and low fuel consumption. This is because it is the excavator's arm and not the excavator that does most of the moving; when they do move, they tend to simply pivot around a ball bearing. Maximum ground speed for an excavator is approximately 3 miles per hour—significantly less than a dozer's 7 miles per hour or wheel loader's 25.

Right: The largest excavator being used for wrecking in the United States today weighs 275,000 pounds without attachments and can reach 60 feet high.

OPERATING AN EXCAVATOR

How do you a control an excavator? With both hands, and feet. The operator's right hand, through the use of a joystick, controls the bucket (attachment) and the boom. Moving the joystick back and forth moves the boom up and down; moving it left and right rotates the bucket around its pivot. The operator's left hand operates the "stick" control and the swing motor that rotates the whole upper structure of the excavator to the left or right. Feet are used to feed hydraulic pressure to the track motors on the undercarriage and propel the excavator.

Below: A chemical plant in Germany being demolished in 1998.

ULTRA-HIGH DEMOLITION: THE CAT 345B

Caterpillar's Ultra-High Demolition 345B excavator (left) allows operators to reach as high as 75 feet. In some instances the excavator is replacing the traditional wrecking ball, because it allows the operator to take a building down in increments, reducing the amount of dust and the danger of collapse. MCM Management Corp., a Detroit-based firm, used the Cat 345B to take down an eight-story building during an emergency job in downtown Detroit; the building was located just 12 feet from an occupied residence and surrounded on all sides with power lines. Using the Cat 345B, MCM was able to take down the 108,000 square foot building in four days—instead of the ten estimated for demolition by wrecking ball.

Front-end Loaders

Front-end loaders can be either tracked or wheeled. Tracked front-end loaders are similar in appearance to bulldozers, with a large, mobile bucket instead of a blade. They are indispensable members of the demolition equipment team, following the excavator to clean the mess left in its wake. The huge bucket scoops up the rubble, which is usually deposited in debris trucks to be removed from the site.

Wheeled loaders can reach speeds of up to 30 miles per hour; tracked can reach only 6. Though the wheeled version is faster, the tracked version is preferred for demolition work—the conditions on demolition sites destroy the wheels almost as quickly as they can be replaced.

Cleanup on a demolition site is a continuous process. Front-end loaders are kept busy separating material for recycling (steel rebar, pipe, I beams, copper wire, electrical motors and aluminum sheet), and for landfills (concrete, brick and wood rubble). In general, the rubble is crushed under the loader's tracks and then scooped up. The rubble is then transferred to another front-end loader for delivery to a debris truck. A typical track loader (like the Cat 935B) weighs approximately 33,000 pounds and can manage a bucket capacity from 2 to 2.4 cubic yards.

Traditional demolition techniques are often used in tight quarters like New York City, where the loaders above are shown clearing sites.

THE WORK OF A LOADER IS NEVER DONE

The loader's routine begins with (not surprisingly) loading. The operator drives the loader up to a pile of rubble, positioning the bucket so it is square to the face of the pile, with the bottom of the bucket just above the ground and tilted slightly forward. The operator then drives the machine forward, rotating the bucket at the same time to scoop up material from the pile. Ideally, the operator will fill 60 to 100 percent of the bucket, depending on the type of rubble. The next task is to maneuver the loader, which generally involves four changes of direction—two at each end of travel. Maneuvering then transitions into travel as the operator navigates the loader from the rubble pile to the "target"—the site, either a truck or a debris pile, where the rubble will be unloaded. Finally, it's time for the loader to dump. The operator, having raised the bucket to the carry position during the maneuvering and traveling stages, is ready to dump as soon as the target is in sight. Taken together, all the steps in the cycle add up to approximately 35 seconds, or 103 cycles per hour.

Loaders like these (left and below) can handle tons of rubble with ease.

Attachments

Attachments (like the shears shown in the pictures above) make it possible to use the same vehicle for a variety of purposes.

MSD 70

⚠ DANGER
KEEP AWAY
75 feet (23m)

SHEARS

Shears are exactly that—monstrous scissorlike attachments that slice quickly and easily through concrete, structural steel, sheet metal, steel plate and just about anything else. When selecting shears, operators must consider the shears' cycle time (how long does it take for the jaw to fully open and close?) and cutting force (how strong are they?).

"The T. rex of mobile demolition shears"
LaBounty's MSD-III Mobile Shears (left), specifically designed for demolition and scrap-metal processing, have been used around the world on such notable demolition projects as the Seattle Kingdome demolition, Eastern European nuclear power plant demolitions, and the Stapleton Airport construction project.

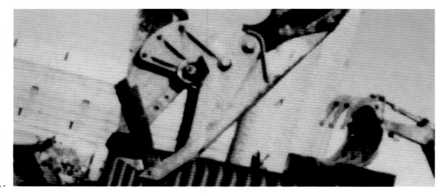

1.

2.

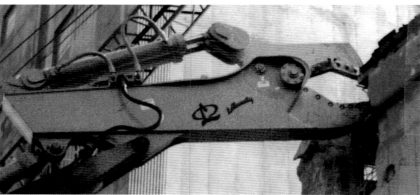

3.

4.

LaBounty Shears over the years
1. 1976—mobile shear to cut steel 2. 1981—shears with increased power
3. 1984—rotating shear 4. 1993—shears with lighter design

ON GRADE PROCESSORS

On Grade Processors were developed to process concrete objects—specifically slabs—located on the ground. The back, or lower jaw, faces the operator to allow better visibility of material processing and easy placement beneath the concrete slabs. Like the Concrete Pulverizer, On Grade Processors excel at separating concrete from rebar.

GRAPPLES

Grapples are (below, right) used in place of a bucket. When attached to an excavator, a grapple allows the operator to grab—at a building, pieces of rubble or anything else. Skilled operators can manipulate the grapple's teeth with pinpoint precision, grabbing at the exact piece of the building they desire.

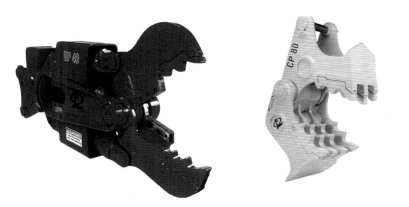

UNIVERSAL PROCESSORS

Universal Processors (above, left) have interchangeable jaws that allow for fast processing of concrete and metal for interior and exterior demolition.

This Universal Processor comes with up to six different sets of jaws—Concrete Cracking Jaws, Concrete Pulverizer Jaws, Shear Jaws, Plate Shear Jaws, Demolition Jaws and Wood Shear Jaws. Jaw sets can be changed in as little as 20 minutes.

CONCRETE PULVERIZERS

Concrete Pulverizers are designed for quiet, controlled demolition and recycling of concrete bridge decks, walls, floor slabs, foundations, silos, culverts, pillars, encased beams, precast structures and railings. Concrete Pulverizers also separate concrete from rebar, producing two recyclable products

The large opening of the Concrete Cracking Jaws is ideal for breaking large concrete structures like parapets, pillars and beams.

Concrete Pulverizer Jaws can be used as a primary demolition tool, separating concrete and rebar as it works, or as a secondary tool, processing material once it is on the ground.

Shear Jaws, though not designed for full-time scrap-metal processing, are excellent for processing rebar structural steel and other types of steel found on demolition sites.

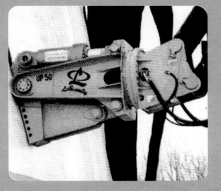

Plate Shear Jaws are for processing storage tanks, be they above or below ground. Because the Jaw cuts plate cleanly, the amount of material distortion is reduced, providing denser loads and greater salvageability.

Possessing the combined features of the Shear Jaw and Concrete Cracking Jaw, the Demolition Jaws can process both heavily reinforced concrete structures and concrete encased I beams.

Wood Shear Jaws are specifically designed for use on stumps, logs and wood demolition debris.

Left: A chimney crashes to the ground after being struck by a wrecking ball.

Bottom right: This excavator, equipped with a hydraulic hammer, works against the backdrop of the Chicago skyline.

Top right: Underwater demolition—a concrete pier is being crunched in a river.

Far right: This excavator is reaching straight for the top of this French apartment building. (Note the different wallpapers of the various apartments.)

CONVENTIONAL DEMOLITION

EBBETS FIELD

LOCATION	Brooklyn, NY
CONTRACTORS	Wrecking Corporation of America
DATE OF DEMOLITION	February, 23, 1960

Ebbets Field (above) hosted hundreds of baseball games prior to its demolition in 1960 (far right).

Ebbets Field, home of the Brooklyn Dodgers, was first opened on April 9, 1913. It hosted its first night game on June 15, 1938, and the final game was played on September 24, 1957. Before moving to their own stadium out west, the Dodgers spent the 1958-1961 seasons at the Los Angeles Memorial Stadium.

Ebbets Field was an elegant structure—the Rotunda was an 80-foot circle enclosed in Italian marble, with a floor tiled with a representation of the stitches of a baseball, and a chandelier with 12 baseball-bat arms holding 12 baseball-shaped globes. There were 12 turnstiles and 12 gilded ticket windows. The domed ceiling measured 27 feet high at its center.

Ebbets Field was the site of two important "firsts": baseball's first televised game (Dodgers versus the Reds) was played on August 26, 1939, and Jackie Robinson became the first African American to play in Major League Baseball on April 15, 1947.

Built on the site of the Pigtown garbage dump for a cost of $750,000, Ebbets Field's demise began on February 23, 1960. Ironically, the demolition company used the same wrecking ball on Ebbets Field that would be used four years later to demolish the Polo Grounds.

The Ebbets Field apartments, built in 1963, now occupy the site.

POLO GROUNDS

LOCATION	New York, NY
CONTRACTORS	Wrecking Corporation of America
DATE OF DEMOLITION	April 10, 1964

As shown at right, fans entered and exited the Polo Grounds through an opening in center field.

The Polo Grounds—filled with fans (below), and reduced to rubble (bottom right).

Polo Grounds served as home to the New York Giants (1911-1957), New York Yankees (1913-1922), and New York Mets (1962-1963). It was officially opened on June 28, 1911, hosted its first night game on May 24, 1940 and its last game on September 18, 1963. Bobby Thomson's "Shot Heard Round the World" home run occurred here at 4:11 PM on October 3, 1951, and won the game for the Giants in the bottom of the ninth inning against the Dodgers in what many consider the greatest game ever played.

The Wrecking Corporation of America, then a joint venture of Cleveland Wrecking Company and Cuyahoga Wrecking Corporation, began demolition of Polo Grounds on April 10, 1964, using the same wrecking ball that had been used on Ebbets Field. Polo Grounds Towers, four 30-story apartment buildings, now occupy the site. The former Polo Grounds center field is now a playground, known as Willie Mays Field; a brass marker indicates the historic significance of the site.

Polo was played in the 1870s on the original Polo Grounds— the piece of land just north of Central Park, bound by 5th and 6th Avenues and 110th and 112th Streets—until it was displaced by baseball in 1883. The New Yorks (later called the Giants) of the National League played in the southeast corner of the park; the Metropolitans, of the American Association, played in the southwest. The Giants were evicted in 1889 and moved uptown to a stadium on the southern end of Coogan's Hollow.

Originally called Manhattan Field, the park was soon dubbed the New Polo Grounds. The Giants moved again in 1891, this time to a stadium called Brotherhood Park built by the Players League on the northern piece of Coogan's Hollow for the 1890 season. (The Players League collapsed after one season.) The Giants once again renamed their new home the Polo Grounds. The venue burned down on April 14, 1911, but was rebuilt later that year.

Below: With no team left, fans could only cheer on the wrecking ball.

PENNSYLVANIA STATION

LOCATION: New York, NY
DATE OF DEMOLITION: 1963

Penn Station's magnificent night-time glow (above), and regal columned entrance (right).

Pennsylvania Station, completed in 1910 by the architects McKim, Mead & White, was one of New York's architectural gems. The Seventh Avenue façade was a pillared portico, the General Waiting Room was reminiscent of ancient Roman monuments, and the train concourse enclosed acres of space with vaulted glass and steel. Travelers arriving in New York City were greeted in style.

As the popularity of airlines and interstate highways grew, train travel began to decline. The interior of Penn Station was another strike against it, as it wasn't well suited for travelers, who had to climb a steep set of stairs in order to reach the street. (Escalators hadn't yet been invented when the station was originally built, and by the time they were widely available, they were too expensive for the Railroad Company to purchase.) The Pennsylvania Railroad Company's response to these conditions was to build a more modern building that would be less expensive and easier to maintain. Irving Mitchell Felt, a New York developer, offered to demolish the station and replace it with an office complex and a brand-new Madison Square Garden.

Before the station's demolition, the president of the Pennsylvania Railroad Company wrote a letter to The New York Times asking the question, "Does it make any sense to preserve a building merely as a 'monument'?" Many New Yorkers were devastated by the demolition of the landmark. The loss of Penn Station intensified the fight, already begun by some of the city's elites, to save New York's architectural treasures from the developmental boom of the postwar years. The group lost the battle for Penn Station, but won the war. Out of this conflict, the New York City Landmarks Preservation Commission was born.

HOW DOES A BUILDING BECOME A LANDMARK?

Buildings older than 30 years are eligible for consideration by the New York City Landmarks Preservation Commission, which appraises them for their architectural, historical or cultural merits. The Commission then holds a public hearing, after which they may designate the building a landmark. Designation as a landmark protects the building from demolition and any alterations believed to be harmful by the Commission.

Millions of travelers passed through old Penn Station's arches (right) before the building was methodically demolished (far right) to make way for the new station and Madison Square Garden.

Above and top right: It took more than two years to dismantle the sprawling Sears Warehouse.

SEARS CATALOGUE WAREHOUSE CENTER

LOCATION	Chicago, Illinois
CONTRACTORS	Brandenburg Industrial Service
DATE OF DEMOLITION	1992-1994

Plans were made to demolish the Sears Catalogue Warehouse Center, a 9-story 3-million square foot brick and timber building erected in 1906, in order to make way for construction of 600 new apartments, townhouses, and single-family homes. With the help of a 165-ton crane, over 7.5 million board feet of yellow pine wood beams and 23 million bricks were recycled.

Above: The empty lot after demolition and debris removal.

Above: Approximately 7.5 million board feet of lumber was recycled from the site of the Sears Warehouse.

A8 HIGHWAY BRIDGE

LOCATION	Highway Bridge on the A 8
CONTRACTORS	Max Wild GmbH
	(Berkheim/Illerbachen)
DATE OF DEMOLITION	Winter 1999

Seven excavators and one wheeled loader were employed to demolish a highway bridge on the Ulm-Munich section of the A8 highway. The bridge, which carried the A8 highway across the Danube River and the ICE-railway tracks, was originally constructed in 1937 in two parts, each with two separate lanes. Heavy traffic eventually exceeded the bridge's capacity, prompting construction of a new interstate bridge with three lanes for each direction.

Clearing the bridge was no simple task: Considerable care had to be taken to avoid damaging the Leipheim water reservoir, the ICE-railway tracks, nearby housing and the Danube riverbed. Further complicating matters, work conducted above the Danube could only be done during low tide.

How did they do it? First, the old road surfaces and the superstructure were brought down using excavators, then the embankment arches on either side of the abutment and the bridge arches over the Danube were blasted. Finally, the bridge-arches over the railway tracks were cut in half with a diamond saw. During the night, a mobile crane removed these sections, and an excavator equipped with a concrete nibbler and backhoe attachment crushed and loaded the rubble for recycling.

Demolition of the old bridge was completed by the end of 1999 to allow for construction of the new bridge, to be ready for traffic in the summer of 2001.

Left: Hard at work on the A8 Highway bridge demolition.

MILLER PARK STADIUM

LOCATION	Milwaukee, Wisconsin
CONTRACTORS	Brandenburg Industrial Service
DATE OF DEMOLITION	August-October 1999

Brandenburg Industrial Service responded to an emergency call during the summer of 1999 after a 567-foot crane collapsed during construction of the Brewers' new Miller Park Stadium. "Big Blue," a 2,100-ton crawler crane, was lifting the last panel of a 400 ton, 180' x 100' 3-section roof when the failure occurred. The collapse brought down temporary supports along with two other sections of the 12,000-ton retractable roof. In addition, two other cranes were destroyed, and portions of the stadium's concrete seating suffered considerable damage. Brandenburg was asked to assess the damage and remove the tangled steel and broken concrete while preserving the remaining structure. To accomplish this task, torches and shears were used to disassemble and cut the steel into pieces suitable for removal.

Big Blue's collapse (above) left a tangled mess of steel and cables (left) and several ruined cranes (middle and top right), and damaged a portion of the stadium's seats (right).

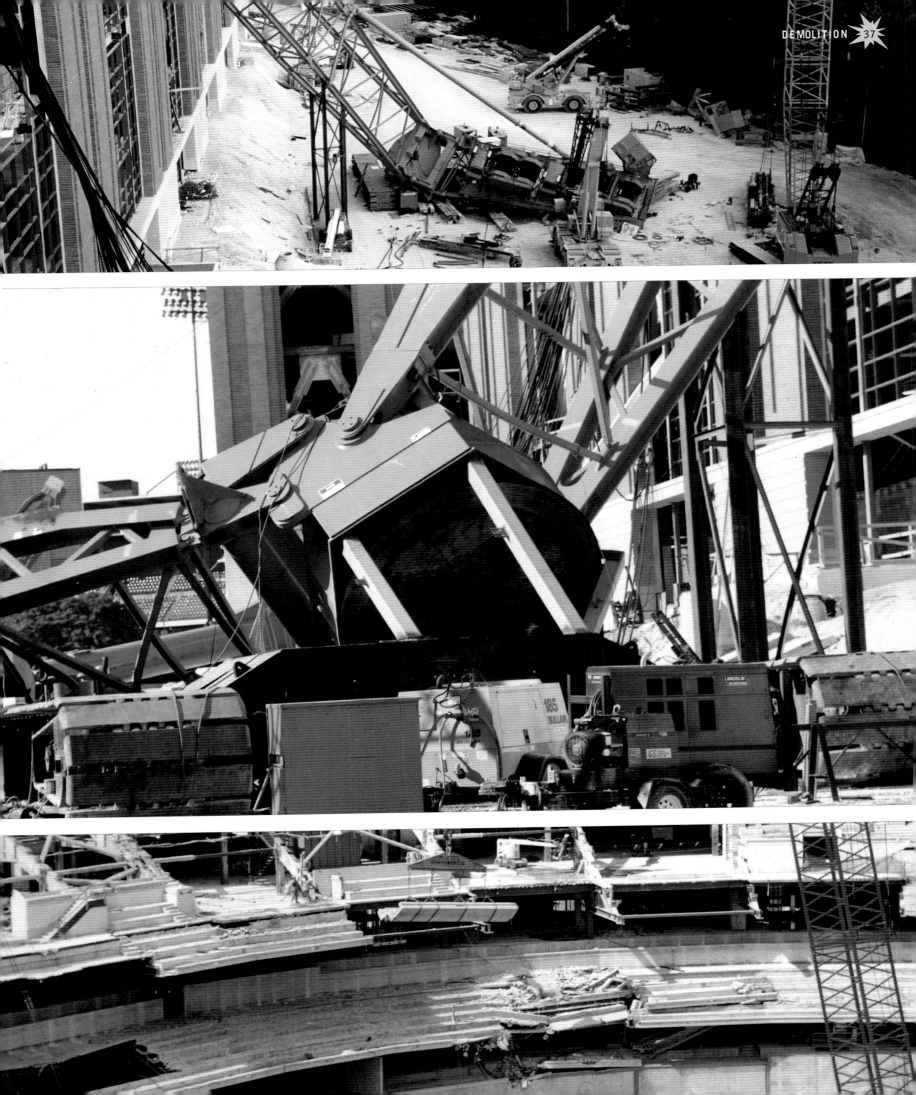

THE EVOLUTION OF IM

THE LOIZEAUX FAMILY OF CONTROLLED DEMOLITION, INC.

Jack Loizeaux first used explosives while a forestry student at the University of Georgia. Responding to the call of a representative from DuPont de Nemours, Jack spent a summer helping to survey land, and lay out, load and detonate several tons of 50% straight dynamite in order to straighten the meandering curves of the Occone River. After graduating, Jack used explosives in the course of tree stump removal and rock blasting. In 1947 he applied explosives to a man-made structure: a 100-foot-tall radial brick chimney. Jack approached the chimney as nothing more than a 'masonry tree,' notching and felling it accordingly, simply substituting explosives in place of a chain saw. Jack's success with this first chimney led to opportunities to bring down a variety of structures.

PLOSION

Jack Loizeaux, founder of Controlled Demolition, Inc.

When people hear the word 'implosion,' they often imagine a building being blown up; the truth, however, is quite the opposite. Through the use of implosion—the strategic placement of explosives that unleash the gravity trapped within—buildings, in fact, fold into themselves, with the rubble often being deposited neatly in the basement. Implosion is not an appropriate method of demolition for all buildings, but for those for which it is appropriate, the result is often swift and awe-inspiring. Onlookers resist blinking—implosions rarely exceed 30 seconds in duration—yet have difficulty believing their eyes. How is a 30-plus-story structure, built with reinforced concrete, reduced to rubble in just mere seconds in such a controlled manner? That which took months or years to build is brought down in seconds. Herein lies the power of implosion.

Left: A bridge implosion performed in Iowa in 1973.

Jack Loizeaux perfected his method, and Controlled Demolition, Inc. was born. His late wife, Freddie, handled all the public relations and is credited with coining the application of the term "implosion." Jack and Freddie's children were loosely involved in the business at early ages, accompanying their father to job sites on weekends and summer holidays. In a freak accident, a car that had skidded on ice struck Jack, who suffered a broken back as a result. Jack's elder son, Mark, stepped up to the plate, leaving college in order to run the company in his father's absence. Mark became the youngest licensed blaster in the United States at the tender age of 20. After his recovery, Jack returned to the helm of CDI, and Mark returned to complete his studies.

According to Mark Loizeaux (right), elder son of Jack and chairman of the Loizeaux Group of Companies, "that which goes up, shall come down."

Below: Freddie Loizeaux, late wife of Jack, was an integral part of CDI's creation.

"Everything vertical to the horizon wants to fall. We just help it." Doug Loizeaux (right), Vice President of CDI, is pictured here with another member of the extended CDI family.

Jack has since retired, and CDI is now run by Mark and Doug (Jack's younger son who joined CDI in 1972), along with Mark's daughter, Stacey. In the course of its 50-plus-year existence, Controlled Demolition has imploded more than 7,000 structures. Their expertise runs the gamut from traditional high-rise buildings to chimneys, bridges, off-shore structures and beyond. Frequently they have come to the aid of earthquake- or bomb-ravaged structures whose instability required deft skillsmanship. Hollywood has also come to appreciate the work of Controlled Demolition, sometimes by filming an already-contracted implosion, and other times by commissioning them to bring down buildings laced with fiery pyrotechnics.

Left: Stacey Loizeaux is CDI's resident pyrotechnic expert, often working closely with Fireworks by Grucci and Hollywood special effects coordinators.

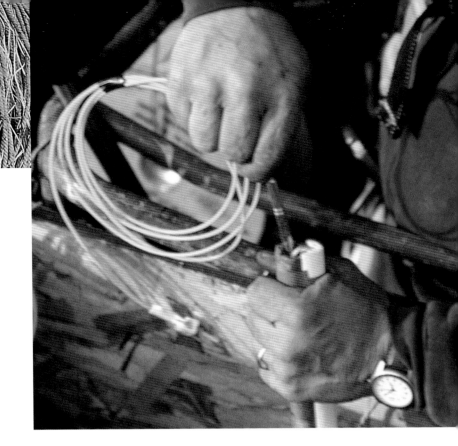

An explosive is a substance or device that is capable of producing a volume of rapidly expanding gas that exerts sudden pressure on its surroundings. There are three principal types of explosives: chemical, mechanical and nuclear. Mechanical explosives create physical reactions, such as when a container is overloaded with compressed air. Nuclear explosives are the most powerful, producing a sustained nuclear reaction, but are generally restricted for use as military weapons. Chemical explosives, which include black powder, nitroglycerin, dynamite and trinitrotoluene (TNT), are the most commonly used. Chemical explosives can be gaseous, liquid or solid—the latter two producing the most powerful results. Chemical explosives are either detonating (high) or deflagrating (low). Detonating explosives, such as dynamite, decompose rapidly and create high pressure. Deflagrating explosives, which may burn quickly, produce considerably less pressure. Two types of detonating explosives exist—primary, detonated by ignition such as a flame or heat-producing impact, and secondary, which require a separate detonator.

HISTORY OF EXPLOSIVES

Explosives powered Mongolian rocket batteries in the 13th century.

Mine blasting, c.1873.

Early powder fuses

Right: Ascanio Sobrero, discoverer of nitroglycerine

Far right: Miners loading fuses

Bottom: A stick of dynamite

BLACK POWDER

The first chemical explosive used was black powder, which was invented in China more than 1,000 years ago. Black powder, a mixture of saltpeter (potassium nitrate), sulfur and charcoal, was originally used strictly for military purposes. A new use for black powder—blasting out mines—was developed in Europe in the 17th century. A flame or intense heat is necessary to detonate black powder. The first detonating systems were often thin, trailing lines of the powder itself or wicks made of straw or other combustible materials sprinkled with powder.

NITROGLYCERIN

Nitroglycerin, discovered in 1846 by the Italian chemist Ascanio Sobrero, and dynamite, invented by Swedish scientist Alfred Nobel in 1867, succeeded black powder as chief explosives. Dynamite was originally a mixture of 75 percent nitroglycerin and 25 percent guhr (a porous, absorbent material that made the product easier to control and safer to use). Nobel developed gelatinous dynamite in 1875, which proved to be more powerful than straight dynamite and safer. Later, ammonium nitrate was used in dynamite, which further increased its safety and decreased its cost.

The first safety fuse was developed by William Bickford of England in 1831; originally a textile-wrapped cord with a black-powder core, the safety fuse allowed for detonations that were both safe and accurately timed. In 1865 Nobel invented the blasting cap, which provided safe and dependable means for detonating nitroglycerin. Electrical firing, which allows greater control over timing, was first employed successfully in the late 19th century.

Left: Detonating cord is often brightly colored—so as not to be missed!

DETONATING CORD

Detonating cord, also known as detonating fuse, contains a high explosive instead of black powder. The first successful cord, patented in France in 1908, consisted of a lead tube filled with TNT that traveled at a velocity of 16,000 feet per second. In 1936 the Ensign-Bickford Company developed Primacord, based on French patents but constructed with different materials, that exhibited a velocity of 21,000 feet per second.

In blasting, a detonating cord is used to connect holes in any pattern desired. The cord is attached to a blasting charge and is fired by either a fuse-type or electric blasting cap. Delay connectors may be inserted to create a delay system.

DELAY SYSTEMS

Delay shooting—separating detonations by as little as a fraction of a second—allows for more precise fragmentation, reduced vibrations, more efficient use of explosives and better control overall.

**Below, top: Wooden boxes perform the same task as sandbags.
Below, bottom: The result of a detonated shaped charge.**

SHAPED CHARGES

Shaped charges, introduced during World War II, normally consist of a metal or glass cone surrounded by a high-strength, high-density explosive. When detonated, the charges produce a high-velocity linear jet blast that can slice through inches of steel quickly and cleanly. Shaped charges were used on the Saturn V moon project to ensure that the spent rocket stages would be cut loose.

Generally, structures fall into two categories—steel and concrete/masonry. Linear-shaped charges are generally used for steel. Linear-shaped charges focus the energy of the charge into a line, generating about 3,000,000 pounds per square inch of pressure. This pressure creates a flow in the steel, forcing the steel aside. Such charges can be used to slice steel as thick as 10 inches. In masonry structures, nitroglycerin- or gel-based stick explosives are generally used.

Right: Sandbags help ensure that the force of the shaped charge is directed appropriately.

Above: Stacey Loizeaux, of Controlled Demolition, Inc., ready to flip the switch on a blasting machine. At the age of three, it was her first implosion.

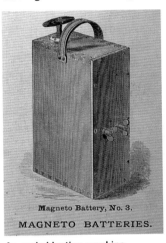

Magneto Battery, No. 3.
MAGNETO BATTERIES.

An early blasting machine.

BLASTING MACHINES

H. Julius Smith invented the first satisfactory electrical blasting machine in 1878. It was comprised of a gear-type arrangement of rack bar and pinion that operated an armature to generate electricity. When the rack bar was depressed rapidly, it revolved the pinion and armature with sufficient speed to create the desired amount of current. This current was released into the external, or cap, circuit when the rack bar struck a brass spring in the bottom of the machine. Eventually, Smith's blasting machine was improved and made in a range of capacities. Though machines like Smith's are still in use, power firing and capacitor-discharge blasting machines have become increasingly more common. Capacitor-discharge blasting machines have a battery source for energizing one or more capacitors and a safe, dependable means for discharging the stored energy.

WHAT DOES IT TAKE TO BE A BLASTER?

Each state has its own requirements that must be met in order to become a licensed blaster. The Occupational Safety and Health Administration also spells out rules in section 1926.901: Blaster qualifications.

(a) A blaster shall be able to understand and give written and oral orders.
(b) A blaster shall be in good physical condition and not be addicted to narcotics, intoxicants or similar types of drugs.
(c) A blaster shall be qualified, by reason of training, knowledge or experience, in the field of transporting, storing, handling and use of explosives, and have a working knowledge of State and local laws and regulations which pertain to explosives.
(d) Blasters shall be required to furnish satisfactory evidence of competency in handling explosives and performing in a safe manner the type of blasting that will be required.
(e) The blaster shall be knowledgeable and competent in the use of each type of blasting method.

THE PROCESS OF IMPLOSION

Implosion is not about blowing up a building, rather it is the process of strategically weakening the structure so it collapses or folds in upon itself. Weeks and often months of preparation are required for a successful implosion. First conventional demolition crews begin the work of emptying, and sometimes weakening, the structure.

During this time, representatives from the implosion team visit and inspect the structure, often with the aid of architectural drawings, to try and understand the true nature of the building. As the building is stripped, the implosion crew begins creating its own drawings and calculations, like assessing the load-bearing walls and figuring out the correct amount and type of explosives to use in order to complete the job. Other issues must be considered as well, like what structures surround the building and what other constraints exist. Next the crew assesses the number and types of holes to be drilled to house the explosives. Test blasts are conducted on columns wrapped in protective material to contain the debris. These test blasts allow the team to assess the strength needed to fracture or eliminate the column. Once this is determined, the drilling begins, followed by the insertion of the charges that will eventually ignite the explosives.

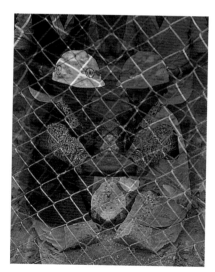

LOADING THE BUILDING

"Loading" refers to the placement of sticks of dynamite or other high explosives into holes drilled into selected columns and load-bearing walls. The shock needed to detonate the explosive is provided by blast-

ing caps, which are placed one to each stick of explosive and set off electrically. When loading dynamite, the blasting cap is pushed into a prepunched hole in the stick of dynamite. The end of the cap—the explosive—part is buried deep inside the dynamite, with detonator wires trailing out behind. When loading a deep hole in a concrete column, the stick is placed in the hole and gently pushed in with a wooden rod. After being situated snugly within the hole, long thin bags of sand, clay or foam are used to fill the rest of the hole, and acts as concrete during the blast, helping to contain the blast and guiding the force to where it is needed in order to split the column open.

HOW MUCH EXPLOSIVES TO USE?

Test blasts are used on columns to determine exactly the right amount of explosives needed to complete the job. The most desirable amount is always the least amount necessary.

Prior to test blasts, the concrete columns are wrapped to contain flying debris. On the left, a column blown with the correct amount of explosives; on the right, a column blown with an excessive amount of explosives.

Implosion is more than just strategic placement of explosives—it also requires careful ordering and timing of these explosives. In order to achieve this, the charges are connected with explosive cord. The first charge is set, and the rest follow in a carefully orchestrated manner. Once the building has been completely loaded and prepared, the crew sets the stage for the actual event. Adjacent streets are closed off to traffic, and security guards sweep the grounds to make sure no one is in danger. When the crew is satisfied that all safety precautions have been met, the stage is set. The countdown begins. Observers hold their breath as the counter approaches "one." A series of loud noises are heard, and everyone watches and waits. Demonstrating expert choreography, the building bends and folds in on itself. The onlookers catch their breath, and applause erupts through the crowd. The implosion was successful, and all that remains is a pile of rubble.

Right: Nearby buildings are sometimes wrapped to protect them from loose debris.

Above: When structures lack adequate innate strength, cables are installed to ensure the extreme walls fall in the right direction.

WHERE DOES IT ALL GO?

Before bringing down a building, there must be adequate space to receive the debris. In instances where this space does not exist, a receptacle must be created.

WHAT ABOUT THE AIR?

Buildings are approximately 85 percent air. When the floors of a building come crashing down, the air is compressed, resulting in the creation of a tremendous blast. It is often this blast of air that is responsible for breaking windows of neighboring buildings. Additionally, loose particles and pieces may be shot out of the breaking columns if the proper protection is not provided—which is why the columns are often wrapped with protective material.

CAN ANYTHING BE IMPLODED?

No! In order to be successfully imploded, a building must be at least five stories tall; anything less will not be heavy enough to allow gravity to properly do its work.

HOW IT ALL FALLS TOGETHER

The direction of the building's collapse is not coincidental. Delay systems are used to knock out supports on one side of the building first—debris will tend to fall in that direction. Another method to ensure the direction of a fall is cabling: Columns are wrapped with steel cable and then secured to a wall; as the columns go down they pull the wall inward, away from the street.

MENDES CALDIERA BUILDING

LOCATION	Sao Paolo, Brazil
CONTRACTORS	Trinton Servicos Especializados e Comercio Ltda (Brazil)/CDI
MATERIAL	Reinforced concrete
DIMENSIONS	32 stories, 361 feet tall
DATE OF DEMOLITION	November 16, 1975

"THAT WHICH GOES UP, SHALL COME DOWN"

Though only 12 years old, the Mendes Caldiera Building was slated for demolition in order to make room for a large station for Sao Paolo's rail-transit system. One thousand pounds of explosives—777 charges—were placed on 11 floors of the structure.

The building proved a challenge for many reasons. One wall of the building was solid, which could have caused the structure to fall sideways rather than crumbling straight down into the 95-foot basement. Also, there were stiff restrictions on where the debris could fall because of the close proximity of the plaza and new station structure already under construction. On two sides, debris had to fall virtually within the lines of the building; on the other two sides, it had to fall within 20 and 33 feet of the lines.

Many were skeptical that the job could be completed safely and successfully: Prior to this building, the highest building ever demolished with explosives was only 22 stories. One local explosives producer refused to bid on the explosives order, for he feared the implosion would end in disaster.

Officials tried to keep the date and time of the implosion a secret, but word of the 7 AM blast on November 16 was leaked. By 3 AM, thousands of onlookers lined the surrounding streets. The blast went off without a hitch at 7:32 AM.

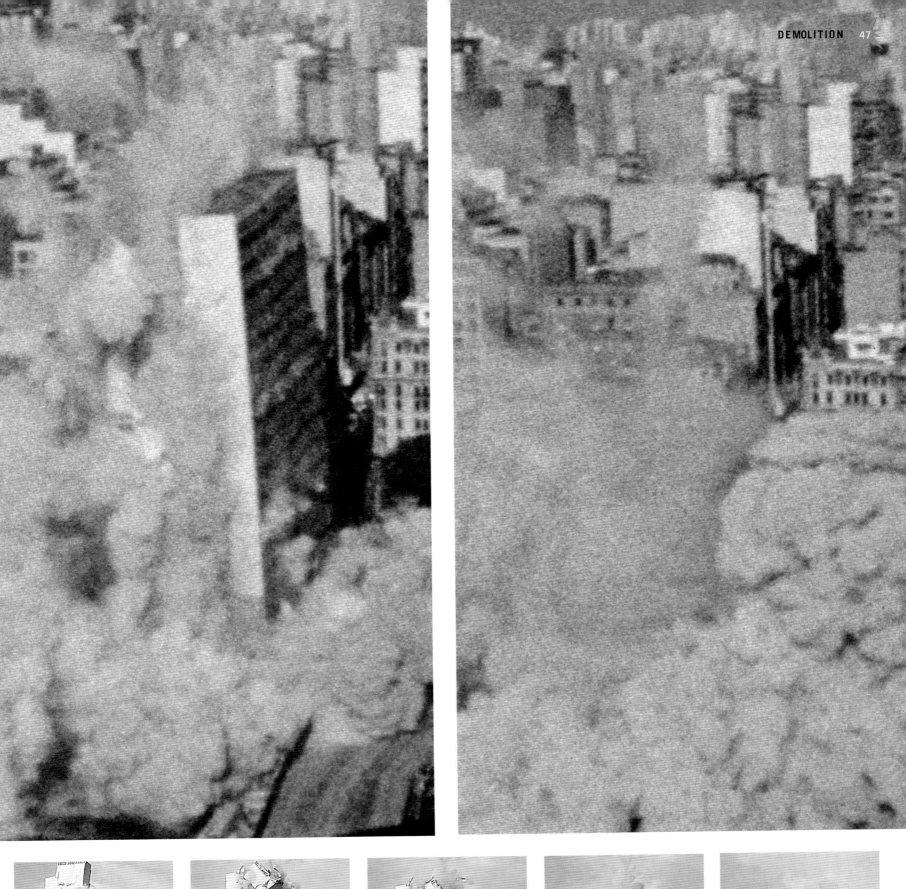

TRAVELER'S BUILDING

LOCATION	Boston, MA
CONTRACTORS	One Twenty Five High Street
	Partnership Ltd. (Boston, MA)/CDI
MATERIAL	Structural steel
DIMENSIONS	18 stories, 451,000 square feet
DATE OF DEMOLITION	March 5, 1988

It took only two weeks to prepare this massive structure for implosion. The building's demolition was challenging—not only was it located in the heart of Boston's congested financial district, adjacent to other buildings, it was also a mere 30 feet from a fire station that could be neither damaged nor impeded in any way. Explosives were placed and detonated strategically to slice the 4-inch thick flanges of the building's support system. (Flanges are protruding edges or rims used to strengthen an object, hold it in place, or attach it to another object.) The building was brought down successfully, and the 60,000 cubic yards of debris were cleared in just six weeks.

ARDLER MSD FLATS

LOCATION	Dundee, Scotland
CONTRACTORS	WJ&D (Contracting), Ltd. (Glasgow, Scotland)/CDI
MATERIAL	Reinforced concrete
DIMENSIONS	18 stories each, 3 buildings
DATE OF DEMO	June 18, 1995

W.J & D (CONTRACTING) LTD
IN ASSOCIATION WITH
C.D.I MARYLAND U.S.A.

Demolition of these 3 18-story buildings was so carefully controlled that all of the debris was contained within the footprint of the structures. (The footprint is the ground that is physically occupied by the structure and outlined by the ground floor curtain walls.)

A worker prepares for the implosion of the Hudson building.

J. L. HUDSON DEPARTMENT STORE

LOCATION	Detroit, MI
CONTRACTORS	Homrich/NASDI, J.V. (Detroit, MI)/CDI
MATERIAL	Structural steel
DIMENSIONS	21 stories, 439 feet tall, 13 structures
DATE OF DEMOLITION	October 24, 1998, 5:47 PM

Prior to its demolition, the J. L. Hudson Department Store was the tallest department store in the United States, second in square footage only to Macy's Herald Square store in New York City. It served the retail needs of Detroit, until closing in 1983.

Demolition of this structure proved a complex task. The store, which had been constructed in 13 stages over the course of 35 years, contained 33 different levels; blueprint drawings were no longer available, further complicating the task of structural analysis. Hudson's was bordered on four sides by streets filled with critical infrastructure and flanked on three sides by poorly maintained turn-of-the-century structures with huge windows that occasionally broke in high winds. Detroit's elevated "People Mover" was located just 15 feet from the structure.

Controlled Demolition, Inc. had to sever the steel in the columns and create a delay system that could simultaneously control the failure of the building's 13 different structural configurations, while trying to keep the hundreds of thousands of tons of debris within the 420 foot by 220 foot footprint of the structure. It took 21 workers three months to investigate the complex and four months to complete preparations for the implosion design.

The columns of the building, each weighing over 500 pounds per foot and containing steel flanges up to $7 \frac{1}{4}$ inch thick, proved yet another challenge. Once steel plates were removed with cutting torches, the remaining steel was cut using smaller charges—a technique intended to reduce the chance of damage to the neighboring windows.

The 12-person loading crew spent 24 days placing 4,118 separate charges in 1,100 locations. Over 36,000 feet of detonating cord and 4,512 nonelectric delay elements were installed in the initiation system to create the 36 primary implosion sequence and 216 micro-delays, which would reduce the pressure created from detonation of the 2,728 pounds of explosives.

Even with all the precautions, seven glass-company crews were alerted to handle any problems. Though over 2,000 yards of soil were placed over utilities in the four adjacent streets, emergency utility crews were still placed on standby.

Nearly 20,000 individuals assembled to watch as the button was pressed at 5:47 PM. When the dust cleared, a debris pile peaking at 60 feet was all that remained.

LARGE PHASED ARRAY RADAR (LPAR) FACILITY

LOCATION	Skrunda, Latvia
CONTRACTORS	U.S. Army Corps of Engineers
	(Winchester, VA)/CDI
MATERIAL	Structural steel and reinforced concrete
DIMENSIONS	Receiver building: 19 stories,
	281 feet tall, 409,300 square feet
DATE OF DEMOLITION	May 4, 1995, 9:45 AM

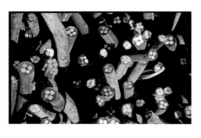

Twenty-five structures, including the second-tallest building in Latvia, were demolished at a Large Phase Array Radar facility in the middle of an active Russian military base in Skrunda, Latvia. The complex was demolished as a result of President Clinton's commitment to help the Latvian government comply with a treaty involving the withdrawal of Russian troops.

A combination of implosion and conventional demolition techniques were employed. Implosion was used to demolish the 19-story receiver building. Conventional methods were used for all other buildings, which included an 8,000-square-foot transmitter building, a transmitter utility

structure, a receiver utility building, a guard house, two warehouses, an ancillary installation building, two transformer substations, a fire station, a neutralizer facility foundation and 25 large underground tanks. Twenty-two thousand pounds of steel, 250 tons of aluminum and 75 tons of copper were recycled and processed for scrap; 5.5 miles of underground cable trench were removed, collapsing the web of underground tunnels that had previously connected the buildings.

Hazardous materials, including asbestos, transformer oil (PCB) and 23 tons of lead battery plates, were removed from the site. The land was then covered with sand to prepare it for agricultural use by local farmers.

Though the project was originally scheduled to take ten months, the work was completed in only five. Six hydraulic excavators, a 100-ton link belt crane, three end-loaders, and four 40-ton tractor trailers were imported from England to facilitate the job. The demolition team included seven Latvian companies, employing as many as 400 Latvian workers during peak times.

The demolition team seized the opportunity to impact the Latvian community in ways other than expected: They brought computers for use in local schools; repaired a church, the City Hall and local waterworks; and provided the community with construction materials and tools to use for future repairs of other structures. Meals were provided for all the local contractors, and training courses were offered and conducted in both Latvian and Russian languages. And after a long day of work, the Americ could often be found playing soccer.

ANZ BANK

LOCATION	Perth, Australia
CONTRACTORS	Mainline Demolition Limited (Perth, Australia)/CDI
MATERIAL	Structural steel and reinforced concrete
DIMENSIONS	14 stories
DATE OF DEMOLITION	July 24, 1991

BARBADOS HILTON HOTEL

LOCATION	St. Michael, Barbados
CONTRACTORS	Polly Septic Services & Equipment Rentals, Ltd. (St. Phillips, Barbados)/CDI
MATERIAL	Reinforced concrete
DIMENSIONS	6 stories
DATE OF DEMOLITION	October 24, 1999, 10 AM

PARKVIEW HILTON HOTEL

LOCATION	Hartford, CT
CONTRACTORS	Manfort Brothers, Inc. (Plainville, CT)/CDI
MATERIAL	Reinforced concrete with a glass curtain wall
DIMENSIONS	17 stories
DATE OF DEMOLITION	October 28, 1990

MONTLUCON COMPLEX

LOCATION	Montlucon, France
CONTRACTORS	Societe Nouvelle de Demolition (Le Plessis, Trevise, France)/CDI
MATERIAL	Structural steel
DIMENSIONS	8 stories, 256,000 square feet, 9 structures
DATE OF DEMOLITION	November 9, 1988

BILTMORE HOTEL

LOCATION	Oklahoma City, OK
CONTRACTORS	Wells Excavating Company (Dallas, TX)/CDI
MATERIAL	Structural steel
DIMENSIONS	28 stories, 245 feet high
DATE OF DEMOLITION	August 28, 1977

The Biltmore Hotel was built in 1932 with heavy beams and reinforced steel columns. By 1977, it stood in the way of a $39-million urban renewal plan to construct a cultural and recreational complex.

The Biltmore presented a challenge because of the heaviness of the steel; each 16-inch steel column with build-up flanges weighed in at 2.5 to 3 tons per floor. A single charge was insufficient to completely penetrate the thickness, so as a result, each 3-inch thick stem plate had to be attacked from both sides. In order to be successful, it was imperative that the charges on the opposing sides go off simultaneously—if one went off too soon, it would dislodge the other before cutting through the steel.

Nine-hundred ninety-one separate charges, approximately 800 pounds of explosives, were placed on seven floors from the basement to the 14th floor and detonated over a five-second span. The building was dropped in a "controlled progressive collapse" that laid it out in such a way as to ease removal of the debris.

OMEGA TOWER

LOCATION	Trelew, Argentina
CONTRACTORS	Tower Inspection, Inc. (Muskogee, OK)/CDI
MATERIAL	Structural steel
DIMENSIONS	1,201 feet tall
DATE OF DEMOLITION	June 23, 1998, 5:40 PM

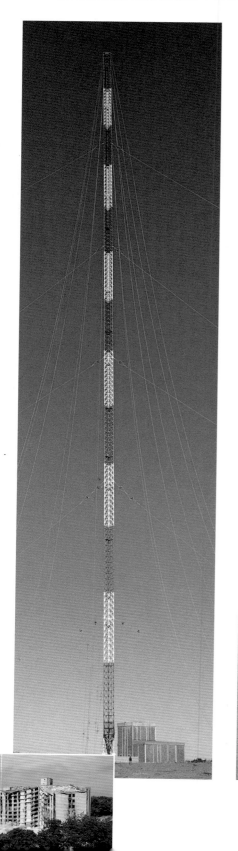

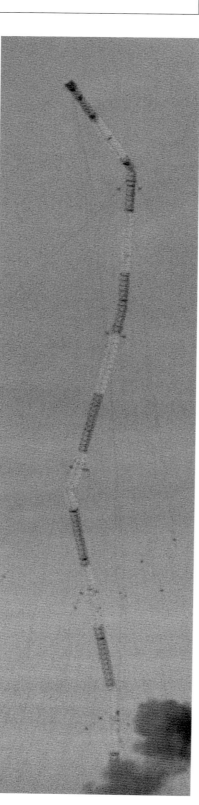

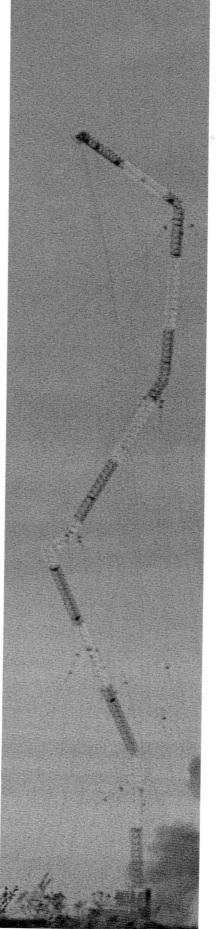

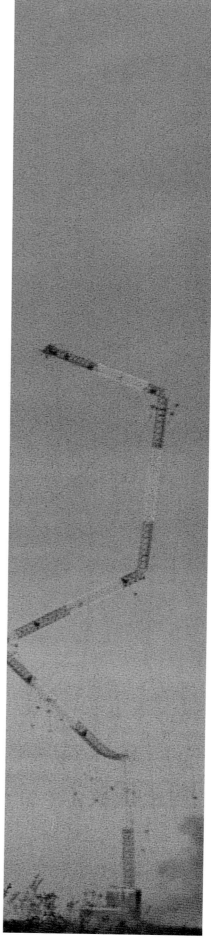

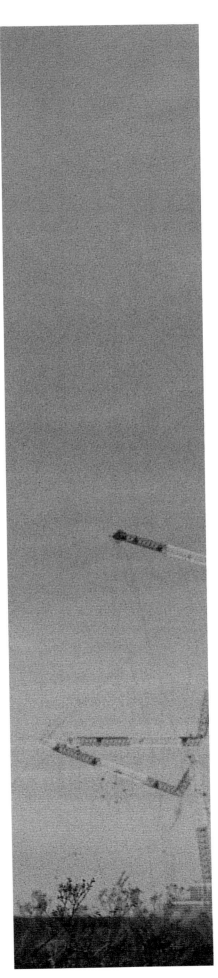

The Omega Tower was felled in an "accordion collapse," into an area less than one-third of its original height and resulted in no damage to the Helix Building located just 21 feet away. The antenna tower and its accessories weighed 310 tons and rested on a ceramic base insulator that supported a total of 1,750,000 pounds. Thirty-four custom-made linear charges, containing 21 pounds of explosives, were used to sever selected structural guys and top-hat radials to initiate the collapse. It took only 11.5 seconds for the tower to fall to the ground, generating vibrations of less than 1 inch per second peak particle velocity—well within the allowable vibrations limits for the safety of the nearby remaining structures.

NAM SAN FOREIGNERS' APARTMENT COMPLEX

LOCATION	Seoul, Korea
CONTRACTORS	Kolon Construction Company, Ltd./CDI
MATERIAL	Reinforced concrete
DIMENSIONS	17 stories, 338,500 square feet, 6 structures
DATE OF DEMOLITION	November 20, 1994

UN PAVILION

LOCATION	Tokyo, Japan
CONTRACTORS	Sumitomo Corporation (Tokyo, Japan)/CDI
MATERIAL	Reinforced concrete, post-tensioned
DATE OF DEMOLITION	March 6, 1986

HOLIDAY INN

LOCATION	Bridgeport, CT
CONTRACTORS	JRP Demolition (Fairfield, CT)/CDI
MATERIAL	Reinforced concrete
DIMENSIONS	10 stories
DATE OF DEMOLITION	May 31, 1998, 7:10 AM

SOUTHWARK TOWERS

LOCATION	Philadelphia, PA
CONTRACTORS	Haines & Kibblehouse, Inc. (Skippack, PA)/CDI
MATERIAL	Reinforced Concrete
DIMENSIONS	2 26-story towers
DATE OF DEMOLITION	January 23, 2000

LAS VEGAS

SANDS HOTEL

LOCATION	Las Vegas, NV
CONTRACTORS	LVI Environmental Services/CDI
MATERIAL	Reinforced concrete
DIMENSIONS	17 stories
DATE OF DEMOLITION	November 26, 1996

The Sands Hotel was only 44 years old when it was demolished in 1996. It took only seven seconds, and seven dynamite blasts, to turn the former playground of the famous Rat Pack into a 30-foot-high pile of rubble.

THE RAT PACK

The Sands was the primary playground of the Rat Pack—a loose group of entertainers that included Frank Sinatra, Dean Martin ("Dag"), Sammy Davis Jr. ("Smokey"), Peter Lawford (JFK's "brother-in-lawford") and Joey Bishop. Shirley MaClaine made occasional appearances with the group as "honorary mascot and Girl Friday."

Legend has it that Lauren Bacall, witnessing her husband Humphrey Bogart returning after a night of carousing with friends such as Frank Sinatra, exclaimed, "You look like a @#$!!! rat pack." The name stuck.

HACIENDA HOTEL

LOCATION	Las Vegas, NV
CONTRACTORS	Goldie Demolishing Company/CDI
	(Las Vegas, NV)
MATERIAL	Concrete Masonry Unit
DIMENSIONS	12 stories
DATE OF DEMOLITION	December 31, 1996 9:00 PM

The Hacienda, noted for its Spanish-style lobby featuring a loud waterfall, was opened in 1957. It grew over the years, spreading over 48 acres and consisting of ten buildings, including "The Little Church of the West," one of Las Vegas's oldest wedding chapels. (Prior to demolition, the chapel was moved from the site.) The Hacienda also served as home to the long-running stand-up comedy show of Redd Foxx, star of the 1970s hit television sitcom "Sanford and Son."

The unique construction of the 11-story, 900-room structure presented some challenges to the implosion crew: The hotel's three wings were

A cloud of dust, set against the spectacular Las Vegas landscape, is all that remains of the Hacienda seconds after its implosion.

built at two different times, under different building codes. The north wing was built in 1980 using concrete block laced with reinforcing rods and filled with grout. The use of precast floor panels created a structure that was stable as long as it remained static. The center tower and south wing were completed in 1989 under more stringent building-code requirements that allowed for greater seismic vibrations. As a result, there was three times more reinforcing in the newer central and south towers than in the original north tower.

Explosives weighing in at 1,125 lb. and 30,600 feet of detonating cord initiating charges in 4,128 different locations brought the three towers down in concert with a massive fireworks display put on by Fireworks by Grucci of Long Island, New York. A four-minute fireworks presentation preceding the implosion of the towers included pyrotechnic waterfalls off the front of the buildings, effects synchronized to music off the roof of the buildings, and a massive aerial display. Following the show, the audience was treated to syncopated patterns of flashes and stripes of fire, created by over 750 pyrotechnic devices and 110 gallons of gasoline placed in the

building. The culmination of these efforts was a 150-foot diameter fireball off the roof of the central tower, which served to ring in the New Year exactly at midnight.

The show was nearly flawless, with the exception of the stairwell at the end of the south tower. The stairwell neglected to fall completely, instead dropping only two stories and listing 10 degrees to the south. Why didn't it fall? Because crews salvaging equipment from the boiler room at that end of the structure had removed load-bearing walls that had been an integral part of the implosion design. It took just 45 minutes on New Year's Day to topple the stairwell with conventional equipment.

The implosion and fireworks extravaganza was carried live on the Fox network. In addition to those who watched it on television, the Clark County Metro Police estimated attendance on the Strip at approximately 600,000.

ALADDIN HOTEL

LOCATION	Las Vegas, NV
CONTRACTORS	LVI Environmental Services/CDI
	(Las Vegas, NV)
MATERIAL	Reinforced concrete
DIMENSIONS	18 stories
DATE OF DEMOLITION	April 27, 1998, 7:30 PM

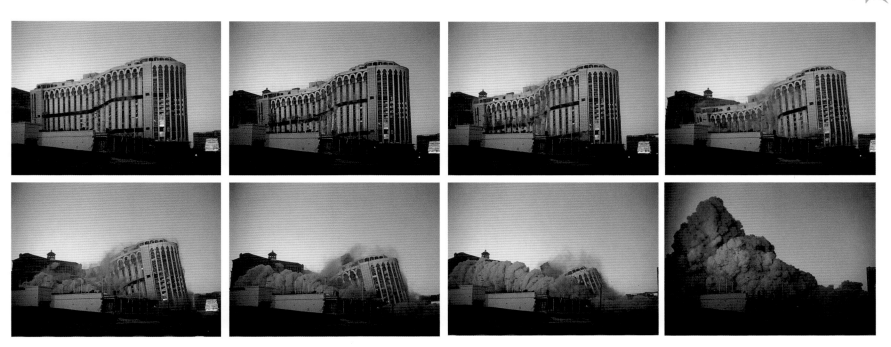

This 1,100-room hotel originally opened in 1963 as the Tally Ho, a nongaming resort catering to wealthy tourists. The Tally Ho failed, and a long list of owners followed, including the legendary Las Vegas performer Wayne Newton. The Aladdin was home to at least one famous wedding—it was here that Elvis and Priscilla Presley tied the knot. And in 1978 the hotel was the site of the world's first slot machine with a jackpot of $1 million.

The Aladdin was leveled in a matter of seconds with 860 charges triggering 600 pounds of gelatin-based dynamite. The charges were placed strategically so the building would fold over itself and spread to the south and the east—away from the Strip. Once the implosion was completed, cleanup crews began sorting through the rubble for recyclable material, including iron, nonferrous metals and concrete. The rest of the rubble was carted away as refuse, a process involving as many as 60 to 70 trucks a day for 20 to 24 days.

WHAT ABOUT THE DUST?

Onlookers were advised to wear dust masks, but county officials were only willing to allow the implosion if the wind speed did not exceed 15 miles an hour.

DUNES HOTEL NORTH TOWER

LOCATION	Las Vegas, NV
CONTRACTORS	Marnell Corrao Associates (Las Vegas, NV)/CDI
MATERIAL	Reinforced concrete and structural steel
DIMENSIONS	22-story building, 18-story sign
DATE OF DEMOLITION	October 27, 1993

The Dunes Hotel, which opened on May 23, 1955, with 194 guest-rooms and the world's largest swimming pool (in the shape of a V), cost $3.5 million to build. The hotel was augmented in 1965 with a 24-story, 250 guest-room North Tower, commonly known as "The Diamond of the Dunes." Steve Wynn, chairman of the Mirage Resorts, Inc., acquired the property in 1992. The following year, Wynn decided to demolish the Dunes through implosion, tying the event to the opening of another of his ventures, Treasure Island hotel and casino. Scenes from the implosion were incorporated into an hour-long feature Treasure Island, the Adventure Begins which aired on national television in January 1994. According to Mr. Wynn, "Anything worth doing is worth overdoing."

The 22-story, 253,000-square-foot reinforced concrete Dunes Hotel was brought down at 10:12 PM on October 27, 1993, in front of more than 250,000 onlookers. The event began with a six-minute fireworks display orchestrated by the famous Grucci family, followed by a pyrotechnic extravaganza choreographed by Controlled Demolition, Inc. First, the 18-story structural steel Dunes' marquee exploded in a shower of sparks and then dove into a section of the hotel. This was followed by 30 seconds of brilliant white flashes and scorching flames. As the smoke cleared, onlookers saw red flashes pulsing quietly in the building as flames flickered throughout. The finale involved the detonation of carefully timed charges, and the Dunes was finally laid to rest.

This incredible show required approximately 364 pounds of explosives, 300 special-effects charges and 460 gallons of aviation fuel. The fireworks component, billed as the largest fireworks display west of the Mississippi, required over 600 man-hours and more than 25 miles of wire to circuit the firing batteries. For their portion of the extravaganza, the Gruccis used 46 firing batteries, 28 tons of sand to fill the batteries and enough board feet of lumber to side a single-family home.

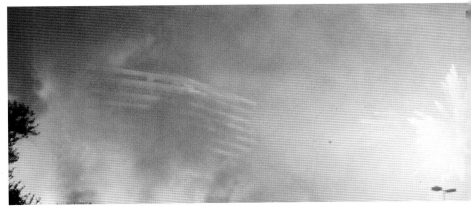

HOLLYWOOD

DR PEPPER BUILDING

LOCATION	Baltimore, MD
CONTRACTORS	No Such Productions, Inc. (Burbank, CA)/CDI
MATERIAL	Reinforced concrete
DIMENSIONS	1-, 3- and 5-story complex, 3 structures
DATE OF DEMOLITION	December 7, 1997

Footage of the Dr Pepper Building implosion was included in the film *Enemy of the State*.

THE DR PEPPER STORY

The Dr Pepper Company is the oldest manufacturer of soft drinks in the United States. Dr Pepper soda pop was first made and sold in 1885 at W.B. Morrison's drugstore in the central Texas town of Waco. Morrison had moved to Texas from Virginia, having worked as a pharmacist in a drugstore in Rural Retreat owned by Dr. Charles Pepper. Charles Alderton, a young English pharmacist working at Morrison's store in Waco, also served carbonated drinks at the soda fountain and is credited with developing the flavor. The name—Dr Pepper—was probably chosen to bolster the image of the drink, which was marketed as a brain tonic and energizer. Morrison began to mass-produce Dr Pepper together with Robert S. Lazenby, who was proprietor of The Circle "A" Ginger Ale Company in Waco. In 1904 Lazenby and his son-in-law, J. B. O'Hara, introduced Dr Pepper to almost 20 million people attending the 1904 World's Fair Exposition in St. Louis. The exposition was the setting for more than one major product debut: Hamburgers and hot dogs were first served on buns at the exposition, and the ice-cream cone was introduced.

ORLANDO CITY HALL

LOCATION	Orlando, FL
CONTRACTORS	Chapman & Sons, Inc. (Orlando, FL)/CDI
MATERIAL	Reinforced concrete
DIMENSIONS	8 stories, 108,000 square feet
DATE OF DEMOLITION	October 25, 1991, 1:31 AM

Implosion of the Old Orlando City Hall was complicated by the New City Hall, which was located just 4 feet away. Twenty million dollars of insurance was furnished to guarantee the safety of the newer civic building.

Warner Bros. Inc., then filming the movie "Lethal Weapon 3", contacted the city fathers of Orlando to see if the City Hall implosion could be utilized as the opening "special-effects sequence" for the new film.

Controlled Demolition, Inc. (CDI) worked with special effects crews in carefully designing an intricate delay sequence that allowed both CDI's explosives and Warner Bros. pyrotechnics to be initiated simultaneously. Through the use of 180 pounds of explosives, in concert with 400 special-effects charges, the 33-year old building collapsed in less than six seconds.

TRAYMORE HOTEL

LOCATION	Atlantic City, NJ
CONTRACTORS	William M. Young & Company, Inc. (Newark, NJ)/CDI
MATERIAL	Reinforced concrete
DIMENSIONS	13 stories, 5 structures
DATE OF DEMOLITION	April 27, 1972; May 26, 1972; August 1, 1972; September 29, 1972

When Atlantic City was founded in 1854, five hotels promptly opened on Atlantic Avenue and featured entertainment in the form of dances, amateur theatrics, concerts and billiards. During the 1870s, more hotels, including the Traymore Hotel, were opened nearer the seaside Boardwalk and amusement piers. Originally designed as a modest cottage, the Traymore kept pace with Atlantic City's growth and reputation, eventually expanding by 1915 to include 600 rooms and a ballroom accommodating 4,000.

The famous director Louis Malle acquired the footage of the Traymore Hotel implosion for use in the opening scenes of his film "Atlantic City."

LANDMARK HOTEL

LOCATION	Las Vegas, NV
CONTRACTORS	Iconco, Inc. (Oakland, CA)/CDI
MATERIAL	Reinforced concrete
DIMENSIONS	31 stories, 77,750 square feet
DATE OF DEMOLITION	November 7, 1995, 5:37 AM

Howard Hughes, the famous billionaire recluse, opened the Landmark Hotel in 1969. Thirty-one stories and 365 feet tall, the structure once held the record for tallest building in the state of Nevada. Constantly beleaguered with financial difficulties, the hotel was closed in 1990 and scheduled for demolition in order to create additional exhibition area for the Las Vegas Visitors and Convention Authority.

The demolition of the Landmark caught the attention of Tim Burton, director of such films as "Edward Scissorhands" and "The Nightmare Before Christmas", who at the time was working on "Mars Attacks!" According to producer Larry Franco, Burton had stayed at the Landmark on a few occasions and was saddened by its imminent destruction.

Burton decided to immortalize the Landmark by incorporating the implosion into "Mars Attacks!" He and his crew were on-site for the implosion, which required 120 pounds of explosives.

Demolition experts make final preparations.

GEORGE P. COLEMAN BRIDGE

LOCATION	Norfolk, VA
CONTRACTORS	Tidewater Construction Company (Norfolk, VA)/CDI
MATERIAL	Structural Steel
DIMENSIONS	2 1,000-foot spans
DATE OF DEMOLITION	July 24, 1996

BRIDGES

The George P. Coleman Bridge is a 3,750-foot-long double-swing-span bridge located in Yorktown, Virginia. Originally built as a two-lane facility in 1952 to connect the counties of York and Gloucester, the bridge supported approximately 15,000 vehicles per day. Due to population growth in the area surrounding the bridge, the load carried by the bridge reached upward of 27,000 vehicles per day by 1995, with no slowdown in sight.

The problem of increased traffic was aggravated by the two-lane structure of the bridge—Route 17 on either side of the bridge had been already enlarged to four lanes to accommodate increased usage. And because it was constructed without shoulders, a broken-down vehicle could cause the lane to be closed down, resulting in hours of frustrating congestion.

Officials recognized the issue of the bridge had to be addressed, but they faced a number of constraints on possible solutions. First, the remedy had to minimize inconvenience on the travelers who relied upon the connection it provided. Second, there were aesthetic concerns from the neighboring historic towns, who did not want their views compromised. And of course there were budgetary constraints.

Working together, the Virginia Department of Transportation and the Tidewater Construction Corporation devised a plan to remove and replace the bridge's spans. The new sections, which were approximately three times wider than the old sections, were built 30 miles downstream and floated up the Chesapeake Bay. Approximately 2,500 feet long, the new sections were floated with everything they'd need to carry traffic—including pavement, light poles and barrier walls. Before the new sec-

tions could be placed, however, the old sections had to be removed. The most effective method of removing the spans, both in terms of dollars and time, was implosion. The implosion team segmented the 4,000-ton bridge into 300-ton sections that matched the lifting capability of Tidewater Construction's equipment. In a matter of seconds the various segments dropped into the Elizabeth River, to be retrieved later. In the end it took only nine days to replace the world's largest double-swing-span bridge.

SUNSHINE SKYWAY BRIDGE

LOCATION	St. Petersburg, FL
CONTRACTORS	The Hardaway Company (Tampa, FL)/CDI
MATERIAL	Structural steel and reinforced concrete
DIMENSIONS	4 main piers, 12 high-level pier, 48 approach piers, 3 anchor spans, 7 SS deck truss spans, 64 footings/seals, 80 RC approaches
DATE OF DEMOLITION	1991-1993

The Sunshine Skyway Bridge straddled Tampa Bay, connecting Bradenton and St. Petersburg. As collisions between oceangoing vessels and the support of the bridge increased, the state decided to replace the bridge with a new, cable-suspension bridge featuring a significantly wider channel. Demolition of the seven deck trusses and four approach thru-trusses and towers on either side of the bridge resulted in bringing down over 9,000 tons of steel. Once the steel was removed, drilling and fragmentation began on 100 pile bents, 48 hammerhead piers, 12 shaft bents and 4 channel piers. The project was designed in such a way that the concrete was fragmented to a 3-foot dimension, eliminating the need for secondary blasting, and in full compliance with the Department of the Environment and Fish and Wildlife regulations to protect the area's endangered species.

MUSCATINE BRIDGE

LOCATION	Muscatine, IA
CONTRACTORS	Dore & Associates Contracting, Inc. (Bay City, MI)/CDI
MATERIAL	Steel
DIMENSIONS	4 spans, 14, 600 feet total
DATE OF DEMOLITION	April 5, April 12, April 13, 1973

MANCHESTER BRIDGE

LOCATION	Pittsburgh, PA
CONTRACTORS	American Bridge – U.S. Steel (Pittsburgh, PA)/CDI
MATERIAL	Structural steel
DIMENSIONS	2 300-foot-long trusses
DATE OF DEMOLITION	September 1970

COURT STREET BRIDGE

LOCATION	Watertown, NY
CONTRACTORS	Joseph B. Fay Company, Inc.
	(Pittsburgh, PA)/CDI
MATERIAL	Reinforced concrete
DIMENSIONS	Two 200-foot arches, 2 abutments
DATE OF DEMOLITION	July 21, 1993

TALMADGE MEMORIAL BRIDGE

LOCATION	Savannah, GA
CONTRACTORS	Anderson Excavating & Wrecking Company (Omaha, NE)/CDI
MATERIAL	Reinforced concrete
DIMENSIONS	1,100-foot span
DATE OF DEMOLITION	December 7, 1992; February 19, 1993

PARKERSBURG-BELPRE SUSPENSION BRIDGE

LOCATION	Parkersbug, WV
CONTRACTORS	Melbourne Brothers Construction Company (North Canton, OH)/CDI
MATERIAL	Steel
DIMENSIONS	2 suspension cables cut at 2 points each, 2 steel towers cut into 5 pieces, and the approach truss into 6
DATE OF DEMOLITION	March 16, 1980

The Parkersburg-Belpre Bridge spanned the Ohio River and connected Parkersbug, West Virginia and Belpre, Ohio. The bridge, 64 years old at the time of demolition, was 2,825 feet long; its main suspension structure had a 775-foot center span flanked by 375- and 275-foot back spans and two 134-feet-high steel towers atop concrete piers rising 35 feet above the river's surface. The bridge also featured 1,400 feet of approach spans.

Melbourne Brothers Construction Company demolished most of the bridge by conventional means—lifting sections out with cranes. Melbourne subcontracted the explosives demolition of the suspension bridge's cables, towers, and a 200-foot truss approach span to Controlled Demolition, Inc. The decision to use explosives was made out of the fear that standard cutting might cause a tower to topple into the channel or the new bridge located just 91 feet away.

The toughness and thickness (8 9/16 inches diameter) of the old suspension cables presented a complex situation: Custom-shaped charges powerful enough to cut the cable without damaging nearby properties could not be developed. In order to overcome this obstacle, the thickness

of the cable was reduced. Clamps on each suspension cable were installed 40 feet from the channel side of each tower. Between the clamps, in two six-inch-long sections, the exterior of each parallel strand wire cable was burned and ground—leaving only 3 inches of thickness instead of 8 9/16 inches.

Shaped charges were then used to cut the cables at the thinned points and drop them, along with the short truss span, into the river in six sections. (These sections were equipped with preattached floats that enabled removal and clearing of the channel in less than three hours.) Relieved of the weight of the main span cables, the towers leaned back toward the shore and away from the channel. Charges were then employed to cut each one vertically through the center and horizontally at two points. As planned, 46 feet of each tower remained standing atop the pier, to be cut apart and removed conventionally.

The new four-lane bridge is a 2,519-foot steel structure with a 1,500-foot cantilever truss.

CHIMNEYS

KENNECOTT COPPER SMELTER STACK

LOCATION	McGill, NV
CONTRACTORS	Kennecott Nevada Copper Company
	(McGill, NV)/CDI
MATERIAL	Reinforced Concrete
DIMENSIONS	750 feet tall
DATE OF DEMOLITION	September 4, 1993

This chimney was the fourth stack to be felled at the Kennecott Corporation's site in McGill, Nevada. Though it measured 750 feet high, 55 feet in diameter, and weighs 9,300 tons, the Kennecott Chimney required only four days' worth of preparation.

HARVARD MEDICAL CHIMNEYS

LOCATION	Boston, MA
CONTRACTORS	Wrecking Corporation of America/CDI
MATERIAL	Radial Brick
DIMENSIONS	2 170-foot towers
DATE OF DEMOLITION	June 26, 1981

Opposite page: Holes were drilled and cables cut prior to the chimney's demolition.

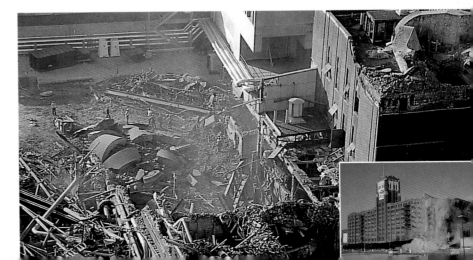

U.S. STEEL FURNACE

LOCATION	Youngstown, OH
CONTRACTORS	Allied Erecting & Dismantling Company (Youngstown, OH)/CDI
MATERIAL	Cast iron and steel
DIMENSIONS	4 structures, weighing 3,000 tons each
DATE OF DEMOLITION	April 28, 1982

INDUSTRIAL STRUCTURES

United States Steel (U.S. Steel) was formed in 1901 by financier John Pierpont (J. P.) Morgan. The corporation was at one time worth $1.4 billion and was comprised of eight major companies, including American Tin Plate Company, American Steel and Wire Company, National Tube Company, American Steel Sheet Company, Carnegie Company, Federal Steel Company and National Steel Company.

U.S. Steel's Ohio Works was originally the Ohio Steel Company, founded in 1892. National Steel Company bought the plant a few years later. National Steel then became part of U.S. Steel, and the Youngstown plant was made part of the Carnegie-Illinois Steel division. The plant was renamed the Ohio Works in 1950.

Four furnaces, each weighing 3,000 tons, were dropped at this site in order to clear the land for future development.

SHARON STEEL PLANT STRUCTURES

LOCATION	Sharon, PA
CONTRACTORS	The Schoonover Company (Ecorse, MI)/CDI
MATERIAL	Steel
DIMENSIONS	3 structures
DATE OF DEMOLITION	Ore Bridge: February 9;
	2 Blast Furnaces: March 10

PSE&G GAS HOLDER

LOCATION	Harrison, NJ
CONTRACTORS	Mercer Wrecking Recycling Corporation (Trenton, NJ)/CDI
MATERIAL	Structural Steel
DIMENSIONS	380 feet tall, 15 million cubic feet capacity
DATE OF DEMOLITION	October 6, 1996, 8:30 AM

MADRAS COOLING TOWERS

LOCATION	Madras, India
CONTRACTORS	M/S Build Aids (Madras, India)/CDI
MATERIAL	Reinforced concrete
DIMENSIONS	3 structures
DATE OF DEMOLITION	February 15, 1997

SEQUEDIN COOLING TOWER

LOCATION	Sequedin, France
CONTRACTORS	Societe Nouvelle de Demolition
	(Les Plessis, Trevis, France)/CDI
MATERIAL	Reinforced concrete
DIMENSIONS	180 feet tall
DATE OF DEMOLITION	November 19, 1982

SPORTS ARENAS

OMNI ARENA

LOCATION	Atlanta, GA
CONTRACTORS	Olshan Demolishing (Houston, TX)/CDI
MATERIAL	Structural Steel
DIMENSIONS	Approximately 150 feet tall
DATE OF DEMOLITION	July 26, 1997, 6:53 AM

Though only 25 years old, the Omni Arena was considered obsolete because it was designed without suites. Turner Sports and Entertainment Development planned to replace the arena with a 20,000-seat multipurpose facility to be completed in time for the 1999-2000 basketball-hockey season.

The Omni Arena was a unique structure, featuring a 360-foot square space-frame roof resting on truss walls that cantilevered (extended outward, supported only at one end). With the assistance of the arena's original structural engineer—and despite the disruption of 13 thunderstorm-triggered site evacuations—the implosion team was able to prepare the structure in just one month.

Design of the implosion was complicated by the close proximity of occupied buildings, a subway and a freight rail line. These structures would not have tolerated the vibrations that would have been created by simply cutting the roof loose and dropping it. So instead of using just 350 charges, 1,500 charges—485 pounds of explosives—were used.

The arena featured a modified folded-plate roof. In profile, there were 25 truncated pyramids ("pods") on a flat roof, which were connected to each other by top-chord steel pipes—connected both diagonally and horizontally on the building's exterior. The roof rested on four perimeter truss walls that were 5 feet thick and 70 feet tall. The 110-foot midsection of the 360-foot wall rested on a buttressed concrete wall, cantilevering 150 feet on one side and 100 feet on the other.

Steel-cutting linear-shaped charges were placed in notches in the pods in order to sever the roof truss system from its supports and drop it to the arena floor 25 feet below entrance level. Within the same sequence, shaped charges were used to demolish the upper sections of the wall trusses. The detonation of the charges was separated by 17 to 35 milliseconds to control the degree of air pressure.

The portion of the arena located a mere 50 feet from the CNN Center, (which housed network's broadcast operations) collapsed first. The collapse then moved across the arena toward the Georgia Dome, 1,000 feet in the opposite direction, in order to direct energy away from CNN. At the same time, additional shaped charges severed selected roof truss elements to "soften" the system, allowing a portion of the energy stored in the roof to be released by the crushing of the frame as it landed. Concrete debris from the structures wrecked earlier was scattered on the floor to absorb energy and reduce vibration.

Earlier in the process, Olshan had gutted approximately 75 percent of the arena's interior to facilitate debris removal by allowing the trusses to scatter over the floor. Gaps in the exterior of the structure had also been created to vent air toward the Georgia Dome. Preparatory work on the frame itself included the removal of secondary wall truss members, cuts in the pod pipes, and the insertion of eight charges per pod. And, because the roof rested on walls through a bearing assembly that did not tie the roof down parallel to the wall trusses, perimeter pods were cable-tied to appropriate walls. This was done to keep the roof from sliding off the walls and helped the pods pull the walls inward during the implosion.

A nonelectric detonation system was selected because of the storm season and electric current in the nearby buildings. On the morning of the blast, two electric blasting caps were hooked up to a capacitor discharge blasting machine located 300 feet away from the arena. One minute before the detonation, the last remaining CNN employee in a nearby control room was evacuated—and allowed to return to the room just three minutes afterward.

ST. LOUIS ARENA

LOCATION	St. Louis, MO
CONTRACTORS	Spirtas Wrecking Company, Inc. (St. Louis, MO)/CDI
MATERIAL	Structural steel and wood
DIMENSIONS	12 stories
DATE OF DEMOLITION	February 27, 1999, 5:45 PM

Designed by a German immigrant and completed in 1929, the St. Louis Arena was structurally complex, featuring arches and trusses made of wood and steel. The nature of the construction—the interdependency of elements and resulting lightweight construction—created a high possibility of uncontrolled and progressive failure if subject to conventional (non-explosive) demolition methods. Implosion appeared to be the solution.

Unlike other implosions, the goal was not to flatten the structure completely—the roof lacked the necessary weight needed to accomplish such a feat. After four weeks of analysis, the implosion sequence was designed: he perimeter of the structure would be dropped 20 feet, and the roof would be dropped to the floor where it could be further demolished by excavators.

One hundred thirty-three pounds—1,462 separate charges—were placed in over 250 locations throughout the Arena. Onlookers were treated to a visual notification of the impending implosion—a pyrotechnic extravaganza designed in concert with Phil Grucci of Fireworks by Grucci. After the show, the explosives were detonated over a period of 14 seconds.

KINGDOME

LOCATION	Seattle, Washington
CONTRACTORS	Aman Environmental Construction Inc. (Oakland, CA)/CDI
MATERIAL	Reinforced concrete
DIMENSIONS	720-feet diameter, 25,000-ton dome
DATE OF DEMOLITION	March 26, 2000

The Kingdome was a multipurpose stadium owned and operated by King County. The 9.1-acre concrete-domed building was home to two major sports teams, trade and consumer shows, big-name concerts and family-oriented activities. Depending upon the event, the stadium could seat up to 70,000 people. Since opening March 27, 1976, close to 3,000 major events have been held, drawing more than 66 million visitors.

In June 1997, Washington State voters approved a statewide referendum to construct a new world-class football/soccer stadium and exhibition center for the community. The new stadium would be home to the Seattle Seahawks, and the new exhibition center would house the many consumer shows that previously took place in the Kingdome. Additionally, a new parking garage would open up more than 2,000 more parking spaces at the facility.

The Kingdome was an engineering marvel, featuring the world's largest thin-shell concrete dome. The roof shell was 5 inches thick, with 40 radial ridges (or ribs) radiating from a 7.5-foot-thick, 28.8-foot-diameter compression ring to a 2-foot deep, 24-foot-wide tension ring. The 2,262-foot-long tension ring, in turn, rested on 40 135-foot-tall perimeter columns.

A crew of 20 prepared the stadium for implosion over the course of five weeks, drilling 5,905 holes for explosives. Finally, 4,728 pounds of explosives and 21.6 miles of detonating cord were used. For the purpose of the implosion, the stadium was divided into six sections. Detonation of the charges was set in phases—three alternate sections in phase one detonated over 3.8 seconds, the remaining sections in phase two, over 6.4 seconds.

The Kingdome was brought to the ground in just 16.8 seconds in a performance that exceeded expectations. Debris from the stadium landed in a 23-foot-high circular pile, instead of the predicted 70 feet. The implosion created the equivalent of a magnitude 2.3 earthquake, whit no vibration damage to adjacent structures.

Cost of original construct: $67 million
Attendance (1976-2000): 73,130,463
Total events (1976-2000): 3,361

Above: Detonating cord dangles in anticipation of the blast.

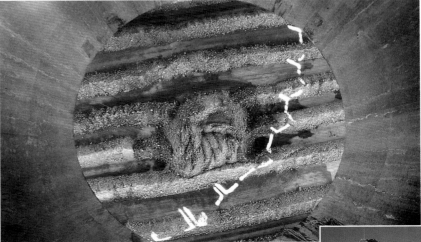

Above; Looking down from a hole at the top of the dome, the cushioning debris is visible hundreds of feet below.

IMPLODING THE KINGDOME — STEP BY STEP

DIAGRAM 1

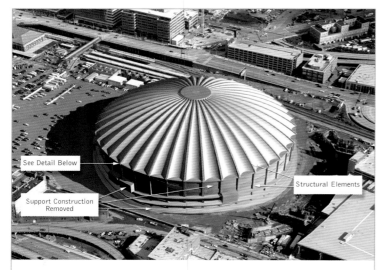

See Detail Below

Support Construction Removed

Structural Elements

The Kingdome consisted of two main categories of construction: Stuctural, which held the building up, and support construction, which could be pre-removed.

The structure consisted of poured-in-place, pre-stressed and post-tensioned concrete elements.

Rebar

Concrete Roof Rib

Concrete Column

DIAGRAM 2

Ramps, Slabs and Siding Removed

Nonstructural ramps, slabs and siding were removed to reduce the amount of debris that fell to the ground.

Explosives were placed in the structure at key locations.

= Explosives

DIAGRAM 3

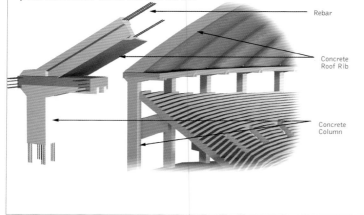

PHASE 2

PHASE 1

Special Wrap

Special Wrap

Demolition of the main structure occurred in two phases of a single implosion sequence. Explosives in the Phase 2 areas were detonated several seconds after Phase 1.

Special material wrapped the structure at blast points to control the explosion.

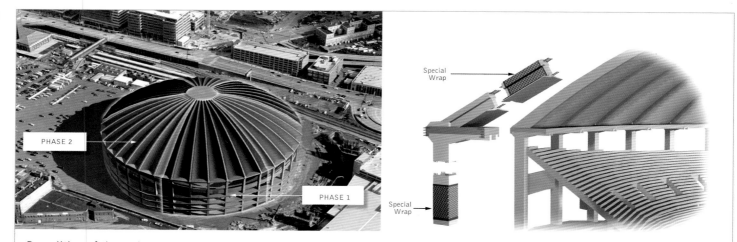

DIAGRAM 4

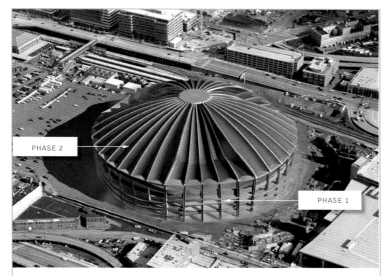

Small explosions fractured the rigid concrete allowing the structure to buckle and fall. The flexible rebar remained intact, acting like ropes to pull the columns toward the center.

Gravity did most of the work.

DIAGRAM 5

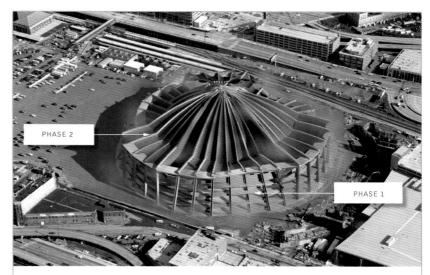

Phase 1 continued to fall, helping to pull Phase 2 in. Phase 2 was detonated several seconds later and collapsed and fell the same as Phase 1.

More explosions further fractured the concrete into smaller pieces.

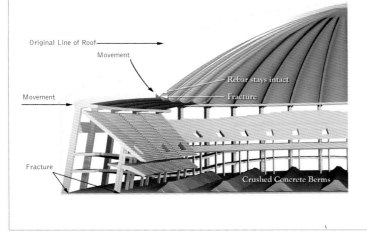

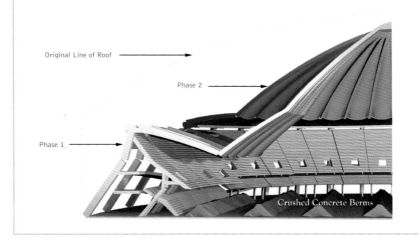

DIAGRAM 6

About 20 seconds after the first explosion, the entire structure collapsed with all of the debris falling within the Kingdome footprint.

The detonation area provided latitude around the entire perimeter for structural debris to fall adjacent to the footprint, but still within the demolition site.

Crushed concrete berms were placed across the Kingdome playing field surface to cushion the fall of the roof. Debris on top of these berms in this area was approximately 25 feet high.

Debris from the perimeter columns and seating construction fell mostly straight down or inward, averaging piles less than 60 feet high.

MILITARY EQUIPMENT DEMOLITION

SCUD MISSILE AND RELATED EQUIPMENT DISPOSAL

LOCATION	Budapest, Hungary
CONTRACTORS	U.S. State Department (Washington, DC)/CDI
DIMENSIONS	7 launchers, 7 fuel tankers, 13 control cabins, 57 nose cones, 59 tail sections
DATE OF DEMOLITION	May 29, 1995

On Monday, May 29, Hungary destroyed its SCUD missile capability before a crowd of dignitaries, which included the minister of defense, U.S. ambassador to Hungary Donald Blinken and a delegation from the United States Congress led by Senator William Roth and Congresswoman Pat Schroeder. The event demonstrated Hungary's commitment to the Missile Technology Control Regime (MTCR), whose purpose is to prevent the proliferation of weapons, and was paid for out of the State Department's nonproliferation fund.

FINISHING NATURE'S WORK

COLONIA ROMA

LOCATION	Mexico City, Mexico
CONTRACTORS	Mexican Government/CDI
MATERIAL	Reinforced Concrete
DATE OF DEMOLITION	1985

At 7:19 AM on September 19, 1985, Mexico City experienced a devastating 8.1-magnitude earthquake. Thirty-six hours later a second earthquake of magnitude 7.5 occurred. The first earthquake shook buildings in Mexico City for a total of three minutes. Unsuspecting citizens were trapped in poorly constructed buildings that had collapsed on them, and many died as a result. The Mexican government estimated some 5,000 people perished, though international agencies placed the death toll at more than 10,000.

The quake destroyed as many as 100,000 housing units and countless public buildings. Government buildings were hit the hardest, especially the Ministries of Communication, Employment, Defense, Education and Urban Development. The epicenter, the point on the Earth's surface directly above the point of rupture, or focus, of the earthquake, was located 50 kilometers (approximately 31 miles) off the coast of Mexico.

The Colonia Roma and Marina Buildings (pages 116-117) were just two of the 26 buildings demolished in response to the earthquake.

MARINA BUILDING

LOCATION	Mexico City, Mexico
CONTRACTORS	Coconal/CDI
MATERIAL	Reinforced concrete; structural steel
DIMENSIONS	1 10-story & 1 12-story structure
DATE OF DEMOLITION	1985

Right: Workers survey the debris left over after the Marina Building was brought down.

EMERGENCY RESPONSE

SHEIK ABDULLAH ALAKL RESIDENTIAL & COMMERCIAL CENTER

LOCATION	Jeddah, Saudi Arabia
CONTRACTORS	Arabian Bechtel Company, Ltd. (Jubail, Saudi Arabia)/CDI
MATERIAL	Reinforced concrete
DIMENSIONS	17 stories, 198,900 square feet
DATE OF DEMOLITION	July 23, 1981

Miscalculation during construction led to the collapse of half of this 12-story building and caused the deaths of 30 construction workers. Responding to an emergency request, the implosion team prepared the structure and successfully demolished it within 96 hours, without causing damage to the adjacent property a mere 8 feet away.

Above: After the initial collapse, the building's supports were cracked and buckling.

Opposite page: Holes were drilled (left) and loaded with explosives (center), then the columns were wrapped prior to implosion (right).

MURRAH FEDERAL BUILDING

LOCATION	Oklahoma City, OK
CONTRACTORS	General Services Administration
	(Fort Worth, TX)/CDI
MATERIAL	Reinforced concrete
DIMENSIONS	9 stories
DATE OF DEMOLITION	May 24, 1995, 7:01 AM

The entire country was shaken on April 19, 1995, by the terrorist attack on the Murrah Federal Building in Oklahoma City. While the nation reeled, Midwestern demolition companies immediately began mobilizing their equipment to the site to assist in the rescue mission. General Services Administration, the owner of the Murrah building, contacted Brockette, Davis & Drake, a structural engineering firm, and Controlled Demolition, Inc., to evaluate the integrity of the structure and identify the extent of demolition required for the portion of the structure that did not fail in the initial collapse. Robert Hill, of BD&D, quickly determined that the main tower of the building was damaged beyond repair, and that the safest way to demolish the building would be through implosion. While the final decisions regarding the fate of the building were being made, the implosion team was contracted to salvage sensitive government paperwork. In addition, granite was salvaged from undamaged portions of the building to be used later in a memorial for the victims of the bombing.

Once the final decision was made to proceed with implosion, preparations began. The three-story underground parking lot to the south of the building was deemed to be structurally sound. All attempts would be made to minimize damage to the structure that was literally touching the Murrah Federal Building.

Preparation for the implosion was extremely complicated. Typically, the contractor relies on the structural integrity of the building being demolished to assist in controlling the fall. In this situation, however, the structural integrity of the building had been compromised. Attempts were made through reconstruction to combat this condition.

Less than 150 pounds of explosives were placed in 420 locations. In just seven seconds, the building, which had been the target of the deadliest domestic terrorist attack in U.S. history, was brought to rest. The American public could begin the next phase of healing.

Right: Flowers were placed around the site in memory of those whose lives were lost.

BIBLIOGRAPHY

DEMOLITION MAGAZINE

ENGINEERING NEWS RECORD

ENCYCLOPAEDIA BRITANNICA

EXPLOSIVES ENGINEERING

Halberstadt, Hans. DEMOLITION EQUIPMENT. Osceola, WI: Motorbooks International Publishers & Wholesalers, 1996.

Loizeaux, J. Mark and Doug K., "Demolition by Implosion." SCIENTIFIC AMERICAN, October 1995.

Norris, Gregory L. "A Total Blast!" HEARTLAND USA, March/April 1999.

Hellman, Hal. "The Demolition Family Bringing Down the House." GEO MAGAZINE.

MODERN MARVELS: DEMOLITION (The History Channel videotape)

Nijkerk, Alfred Arn. and Dalmijn, Wijnand L. HANDBOOK OF RECYCLING TECHNIQUES. The Hague: Nijkerk Counsulting, 1998.

PRESERVING THE PAST, PREPARING THE FUTURE (National Association of Demolition Contractors' videotape)

Robson, Nancy Taylor, "A Booming Business." MARYLAND MAGAZINE, March 1996.

Samuels, David, "Bringing Down the House: An Explosion in Las Vegas Plays as Performance Art." HARPER'S MAGAZINE, July, 1997.

PHOTOGRAPHY CREDITS

INDEX

A

A8 Highway Bridge (Germany), 35

"accordion collapse," 63

air, 45; pressure, 104

Aladdin Hotel (Las Vegas, NV), 74-75

Alderton, Charles, 79

Allied Erecting & Dismantling Company, 98

Aman Environmental Construction, Inc., 108-109

American Bridge-U.S. Steel, 91

Anderson Excavation & Wrecking Company, 94

antenna towers, 62-63

ANZ Bank (Perth, Australia), 56-57

apartment buildings, 50-51, 64

Arabian Bechtel Company, Ltd., 118

Ardler MSD Flats (Dundee, Scotland), 50-51

Atlanta, GA, 104-105

Atlantic City, NJ, 82-83

ATLANTIC CITY (movie), 83

attachments, 24-27; stick-end, 18;
    SEE ALSO concrete pulverizers; grapples; on grade processors;
    shears; universal processors

B

Bacall, Lauren, 71

backhoes, 18, 35

Baltimore, MD, 78-79

banks, 56-57

Barbados Hilton Hotel (St. Michael, Barbados), 56-57

Bastille, 10

Bastille Day, 10

battering rams, 18

Berlin Wall (Germany), 10-11

Bickford, William, 42

"Big Blue" (crane), 36, 37

Biltmore Hotel (Oklahoma City, OK), 60-61

Bishop, Joey, 71,

black powder, 42

blasters, 43

blasting caps, 42, 44, 104

blasting machines, 43, 104

Blinken, Donald, 113

Bogart, Humphrey, 71

Boston, MA, 48-49, 97

Brandenburg Industrial Service, 34, 36

Bridgeport, CT, 66-67

bridges, 35, 88-90, 91, 92-93, 94; double-swing-span, 80-87;
    suspension, 95

Brockette, Davis, & Drake, 120

Brooklyn, NY, 30

Budapest, Hungary, 112-113

building codes, 72-73

building collapses, 114-115, 116-117, 118-119

Burton, Tim, 85

C

cables, suspension, 95

cabling, 45

Callieach, 10

casinos, 71, 74-75, 76-77

Cat 345B, 21

Cat 935B, 22

Caterpillar Ultra-High Demolition 345B, SEE CAT 345B

Chapman & Sons, Inc., 80

Chicago, IL, 34

chimneys, 96, 97

city halls, 80-81

civic buildings, 80-81, 120-121

Clark County Metro Police, 73

Clinton, President Bill, 54

CNN Center (Atlanta, GA), 104

Coconal, S.A., 68

Coliseum (Rome, Italy), 10

Colonia Roma (Mexico City, Mexico), 114-115

columns, 44, 53, 60

concrete, 27, 90, 104

concrete pulverizers, 27

construction accidents, 36, 118

Controlled Demolition, Inc., 39, 40, 41, 53, 76, 80, 95, 120 "controlled progressive collapse," 60

cooling towers, 102, 103

Court Street Bridge (Watertown, NY), 92-93

crane-and-ball, 19

cranes, 16, 35, 95; belt, 55; crawler, 36

D

Danube River, 35

Davis, Jr., Sammy, 71

delay systems, 43, 45, 55, 80

demolition contractors, 15, 17

demolition, process of, 14-17, 54; safety of, 17; technological
          innovation in, 18

department stores, 52-53

detonating cords, 43, 53, 73, 110

detonating fuses, 43

Diamond of the Dunes, SEE Dunes Hotel North Tower (Las Vegas, NV)
          domes, concrete, 108; thin-shell, 108

Dore & Associates Contracting, Inc., 91

Dr. Pepper (soda), 79

Dr. Pepper Building (Baltimore, MD), 78-79

Dr. Pepper Company, 79

Dundee, Scotland, 50-51

Dunes Tower North (Las Vegas, NV), 76-77

dust, 75

dynamite, 42, 44, 71, 75

E

earthquakes, 12, 115

Ebbets Field (Brooklyn, NY), 30

Elizabeth River, 87

ENEMY OF THE STATE (movie), 78

Ensign-Bickford Company, 43

excavators, 20-21, 35, 106; control of, 21; hydraulic, 17, 20, 55;
     SEE ALSO CAT 345B

explosives, 42-43, 73, 76, 80, 85, 95, 104, 106, 108, 120;
     SEE ALSO dynamite

F

factories, 78-79

Felt, Irving Mitchell, 32

fireballs, 73

fires, 12, 73, 76

Fireworks by Grucci, 41, 73, 76, 106

fireworks, 41, 73, 75, 76, 106

flanges, 49, 53, 60

Florida Department of the Environment and Fish and Wildlife, 90

footprints, 51, 53

Fox network, 73

Foxx, Red, 72

Franco, Larry, 85

front-end loaders, 15, 17, 22-23, 35, 55

furnaces, 98-99, 100

fuses, SEE detonating fuses; safety fuses

G

General Services Administration, 120

George P. Coleman Bridge (Yorktown, VA), 86-87

Georgia Dome (Atlanta, GA), 104

Goldie Demolishing Company, 72

granite, 120

grapples, 27

Great Fire of London, SEE London, Great Fire of

Grucci, Phil, 106

Gulf of Mexico, Mexico, 68-69

H

Hacienda Hotel (Las Vegas, NV), 72-73

Hardaway Company, 88

Hartford, CT, 58

Harvard Medical Chimneys (Boston, MA), 97

hazardous materials, removal of, 54

Helix Building (Trelew, Argentina), 63

Hill, Robert, 120

historic preservation, 17, 32, 86

Holiday Inn (Bridgeport, CT), 66-67

Hollywood, CA, 41, 78, 80, 83, 84

Homrich/NASDI, J. V., 52

hotels, 56-57, 58, 60-61, 66-67, 71, 72-73, 74-75, 76-77

houses, 15

Hudson Department Store, SEE J. L. Department Store (Detroit, MI)

Hughes, Howard, 85

hydraulics, 17

I

ICE railway tracks, 35
Iconco, Linc., 84
Imhotep, 10
implosion, 17, 39, 40, 41, 44-45, (images 46-121)
industrial structures, 98-99, 100-101, 102, 103
International Omega Technical Commission, 63

J

J. L. Hudson Department Store (Detroit, MI), 52-53
jaws, concrete cracking, 27; concrete pulverizer, 27; demolition, 27;
        plate shear, 27; shear, 27; wood shear, 27
Jeddah, Saudia Arabia, 118-119
Joseph B. Fay Company, Inc., 93
JRP Demolition, 66

K

Kali, 10
Kennecott Copper smelter stack (McGill, NV), 96
Kennecott Nevada Copper Company, 96
King Charles II, 12
King Djoser, 10
Kingdome (Seattle, WA), 26, 108-111
Kolon Construction Company, Ltd., 64

L

LaBounty MSD-111 Mobile Shears, 26
Landmark Hotel (Las Vegas, NV), 84-85
landmarks, 32
Large Phased Array Radar (LPAR) facility (Skrunda, Latvia), 54-55
Las Vegas Visitors and Convention Authority (Las Vegas, NV), 85
Las Vegas, NV, 71, 72-73, 74-75, 76-77, 84-85
Lawford, Peter, 71

Lazenby, Robert S., 79
LETHAL WEAPON 3 (movie), 80
lettre de cachet, 10
Little Church of the West (Las Vegas, NV), 72
"loading," 44
Loizeaux, Doug, 41
Loizeaux, Freddie, 40
Loizeaux, Jack, 39-41
Loizeaux, Mark, 40
Loizeaux, Stacey, 41
London, Great Fire of, 12
LVJ Environmental Services, 74

M

M/S Build Aids, 102
Madison Square Garden (New York, NY), 32
Madras, India, 102
Madras cooling tower (Madras, India), 102
Mainline Demolition Limited, 56
Malle, Louis, 83
Manchester Bridge (Pittsburgh, PA), 91
Manfort Brothers, Inc., 58
Marina Building (Mexico City, Mexico), 116-117
MARS ATTACKS! (movie), 85
Martin, Dean, 71
Max Wild GmbH, 35
McGill, NV, 96
McKim Mead & White, 32
McLaine, Shirley, 71
MCM Management Corporation, 21
Melbourne Brothers Construction Company, 95
Mendes Caldeira Building (Sao Paulo, Brazil), 46-47
Mexico City, Mexico, 114-115, 116-117
military complexes, 54-55, 112-113
military equipment, 54-55, 112-113
Miller Park Stadium (Milwaukee, WI), 36
Milwaukee, WI, 36
Mirage Resorts, Inc., 76
Missile Technology Control Regime (MTCR), 113
Montlucon Complex (Montlucon, France), 59
Montlucon, France, 59
Morgan, J. P., 98
Mornell Corrao Associates, 76

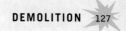

Morrigan, 10

Morris, Charles, 12

Morrison, W. B., 79

"The Monument" (London, England)(column), 12

Munich, Germany, 35

Murrah Federal Building (Oklahoma, OK), 120

Muscatine, IA, 91

Muscatine Bridge (Muscatine, IA), 91

N

Namsam Foreigners' Apartment Complex (Seoul, Korea), 64

National Association of Demolition Contractors, 17

New Year's Eve (1996), 73

New York, NY, 30, 31, 32-33

New York City Landmarks Preservation Commission, 32

Newton, Wayne, 75

nibblers, 35

nitroglycerin, 42

No Such Productions, Inc., 78

Nobel, Alfred, 42

O

O'Hara, J. B., 79

Ohio River, 95

Ohio Works, 98

Oklahoma City, OK, 60-61, 120-121

Olshan Demolishing, 104

Omega Navigational System, 63

Omega Tower (Trelew, Argentina), 62-63

Omni Arena (Atlanta, GA), 104-105

on grade processors, 27

One Twenty Five High Street Partnership, 48

onlookers, 46, 52, 73, 75, 76, 104

Orlando, FL, 80-81

Orlando City Hall (Orlando, FL), 80-81

P

Parkersburg, WV, 95

Parkersburg-Belpre Suspension Bridge (Parkersburg, WV), 95

Parkview Hilton Hotel (Hartford, CT), 58

pavilions, 65

Pennsylvania Railroad Company, 32

Pennsylvania Station (New York, NY), 32-33

Perth, Australia, 56-57

Pittsburgh, PA, 91

"pods," 104

Polly Septic Services & Equipment Rentals Ltd., 56

Polo Grounds (New York, NY), 31

Presley, Elvis, 75

Presley, Priscilla, 75

Primacord, 43

Ptah, 10

R

railway stations, 32-33

Ramos, Andreas, 11

Rat Pack, 71

rebar, 20, 27

recycling, 17, 34, 35, 55, 75

refuse, 75

reinforcing bar, SEE rebar

roofs, lightweight, 106; space-frame, 104

Roth, Senator William, 113

S

safety fuses, 42

San Francisco, CA, 12

Sands Hotel (Las Vegas, NV), 71

Sao Paulo, Brazil, 46-47

Savannah, GA, 94

Schoonover Company, 100

Schroeder, Congresswoman Pat, 113

SCUD missile and related equipment disposal (Budapest, Hungary), 112-113

Sears Catalogue Warehouse Center, (Chicago, IL), 34

Seattle, WA, 108-111

Sekhmet, 10

Seoul, Korea, 64

Sequedin cooling tower (Sequedin, France), 103

Sequedin, France, 103

shaped charges, 43, 95, 104

Sharon Steel Plant (Sharon, PA), 100-101

Sharon, PA, 100-101

shears, 25, 26, 36

Sheik Abdullah Alakl Residential & Commercial Center (Jeddah, Saudi
        Arabia), 118-119

Sinatra, Frank, 71

Skrunda (Latvia), 54-55

Sobrero, Ascanio, 42

Societe Nouvelle de Demolition, 59, 102

special-effects charges, 80

Spirtas Wrecking Company, Inc., 106

sports arenas, 26, 104-105;
        SEE ALSO stadiums

St. Louis, MO, 106-107

St. Louis Arena (St. Louis, MO), 106-107

St. Michael, Barbados, 56-57

St. Paul's Cathedral (London, England), 12

St. Petersburg, FL, 88-90

stadiums, 30, 31, 36;
        SEE ALSO sports arenas

steel, 90

Sumitomo Corporation, 65

Sunshine Skyway Bridge (St. Petersburg, FL), 88-90

T

Tally Ho, SEE Aladdin Hotel

Talmadge Memorial Bridge (Savannah, GA), 94

Tampa Bay, 990

terrorist attacks, 120

Tidewater Construction Company, 86, 87

Tokyo, Japan, 65

torches, 36

Tower Inspection, Inc., 62

Traveler's Building (Boston, MA), 48-49

TREASURE ISLAND, THE ADVENTURE BEGINS (TV show), 76

Trelew, Argentina, 62-63

Trinton Servicos Especializados e Comercio Ltda, 46

truss walls, 104

Tryamore Hotel (Atlantic City, NJ), 82-83

Turner Sports and Entertainment Development, 104

U

U.S. Army Corps of Engineers, 54

U.S. Coast Guard, 63

U.S. Department of Transportation, 17

U.S. Environmental Protection Agency, 17

U.S. Federal Aviation Administration, 63

U.S. Navy, 63

U.S. Occupational Safety and Health Administration, 43

U.S. Office of Safety and Health Agency, 17

U.S. State Department, 113

U.S. Steel furnace (Youngstown, OH), 98-99

U.S. Steel, 98

Ulm, Germany, 35

UN Pavilion (Tokyo, Japan), 65

universal processors, 27

V

Van Ness Avenue (San Francisco, CA), 12

Vatican (Italy), 10

vibrations, 104

Virginia Department of Transportation, 86

W

warehouses, 34

Warner Bros., 80

waterfalls, 72, 73

Watertown, NY 92-93

Wells Excavating Company, 60

William M. Young & Company, Inc., 83

Willie Mays Field (New York, NY), 31

WJ & D (Contracting) Ltd., 50

World's Fair Exposition, 1904 (St. Louis, MO), 79

Worldwide Omega System, 63

wrecking balls, 17, 19, 30, 31

Wrecking Corporation of America, 31, 97

Wynn, Steve, 76

Y

Yorktown, VA, 86-87

Youngstown, OH, 98-99